The birds of Cambridgeshire

Frontispiece: in the early 1950s, thousands of Pinkfeet flighted to the Nene Washes.

THE BIRDS OF CAMBRIDGESHIRE

P. M. M. Bircham

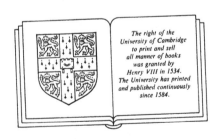

CAMBRIDGE UNIVERSITY PRESS

Cambridge

New York   New Rochelle   Melbourne   Sydney

CAMBRIDGE UNIVERSITY PRESS
Cambridge, New York, Melbourne, Madrid, Cape Town, Singapore, São Paulo, Delhi

Cambridge University Press
The Edinburgh Building, Cambridge CB2 8RU, UK

Published in the United States of America by Cambridge University Press, New York

www.cambridge.org
Information on this title: www.cambridge.org/9780521111720

First published 1989
This digitally printed version 2009

A catalogue record for this publication is available from the British Library

Library of Congress Cataloguing in Publication data
Bircham, P.M.M. (Peter Michael Miles), 1947–
The birds of Cambridgeshire / P.M.M. Bircham.
p. cm.
Includes Index.
ISBN 0 52132863 2
1. Birds – England – Cambridgeshire. I. Title.
QL690.G7B54 1989
598.29426'5–dc 19 88-38628 CIP

ISBN 978-0-521-32863-0 hardback
ISBN 978-0-521-11172-0 paperback

DEDICATION

This book is dedicated to the memory of three early members of the Cambridge Bird Club

Constantine Walter Benson 1909–1982

David Lambert Lack 1910–1973

William Homan Thorpe 1902–1986

CONTENTS

PREFACE

My interest in birds began as a child in the countryside around my home and I was encouraged by the gift of the three volumes of *The Ladybird Book of Birds* which made such birds as Chaffinch, Bullfinch, Willow Warbler, Blackcap, Redwing and Fieldfare all identifiable.

At preparatory school I was most fortunate to be taught Natural History twice a week by Susan Taylor, and at the tender age of eight or nine was shown bird ringing for the first time which I found enthralling. She it was who sowed the seed. However, it was not until my late teens that I resumed my interest. I began colour ringing birds in our vicarage garden and joined the British Trust for Ornithology, yet somehow I remained totally isolated from any other birdwatchers.

When I went to work in the University of Cambridge my interest in birds became known and I was given an introduction to Chris Thorne who lifted the level of my interest almost overnight. He took me to the Cambridge Bird Club, where I remember feeling utterly overawed, and he also gave me tuition in the art of bird ringing so that within a few months of this meeting I found myself in a room full of people I did not know, discussing the formation of a ringing group to be centred on Wicken Fen.

At Wicken, during those long weekends surrounded by bird talk, I learned a great deal but I was still hesitant when asked by Michael Allen, the secretary of the group at that time, to write something for the next Wicken Fen Group Report. Thus began my interest in the status of birds, since I chose to look at a bird whose population was increasing: the Redpoll.

It was 14 years later, in a Canterbury book shop, when I came upon the paperback version of the *Birds of Kent* and bought it as an example of a county avifauna that represented value for money, that I became aware that I could write something comparable for Cambridgeshire. It had long been discussed by the Cambridge Bird Club but I knew that previous attempts had failed so I began work without advertising my project. Half-way through the systematic list I decided to approach Cambridge University Press with a view to publication and to my surprise they expressed interest.

This publication has taken two years to write and follows just over 50 years from the last that bore this title. Chris Thorne and Nick Davies have read the whole manuscript and used their proof-reading skills to correct typographical errors and improve my syntax! Bill Jordan allowed me to use an amended version of his article 'Where to watch birds in Cambridgeshire' and Tony Edwards has watched me with utmost patience using valuable experimental time in the laboratory to read ornithological articles, and my family have supported my interest. To all these people my thanks are due in no small measure.

My greatest debt is to Graham Easy who has given me more help than I had any right to ask and has read, corrected and constantly improved the text of the systematic list. All the remaining imperfections are my responsibility.

Finally, I must thank Martin Walters of Cambridge University Press for all his help and encouragement during the many stages of production.

Peter Bircham
Cambridge 1988

EXPLANATORY NOTE

This book was written to include all the records for the old county of Cambridgeshire up to, and including, the year 1986. Despite the local government re-organisation of 1972, most naturalist societies have retained the old boundaries in order to use comparable data and although a strong case was made to produce a book which included old Huntingdonshire this proved too great a task. Throughout the book the references to Cambridgeshire therefore refer to the old county. Likewise, there may appear references at times to Huntingdonshire, a county which is no longer in existance for government administrative purposes.

Certain abbreviations have been employed for the various organisations mentioned in the text. These are:

BTO–British Trust for Ornithology

CWT–Cambridgeshire Wildlife Trust

RSPB–Royal Society for the Protection of Birds

1
THE CAMBRIDGESHIRE COUNTRYSIDE

Cambridgeshire covers an area of over 500 000 acres most of which is good, not to say rich, agricultural farmland. Surprisingly, the county borders as many as eight others, in clockwise order: Norfolk, Suffolk, Essex, Hertfordshire, Bedfordshire, Huntingdonshire, Northamptonshire and Lincolnshire. At its northernmost point (Wisbech) it is only five miles from the sea, and the River Nene is tidal for a way into Cambridgeshire, yet for all that the lack of coastline is very significant since it restricts both the number of species recorded in the county and the occurrences of those marine birds that wander inland. It is easy to divide the county into north (fenland) and south (chalk and boulder clay) by topography and it is within these divisions, plus suburbia, that the county is described below.

FENLAND

The northern part of the county includes what remains of the vast watery fen that Vermuyden succeeded in draining. It is a flat, somewhat desolate landscape, short on trees and rich in agricultural production, celery, potatoes, carrots and beans are grown in plenty on the black peaty soils. Many a visitor finds the fenland unprepossessing and featureless but to the native and the ornithologist it is an area of immense possibilities. Ever since the Romans came various attempts had been made to drain the marshland which stretched from Waterbeach in the south through to Lincolnshire in the north and Suffolk in the east (see Fig. 1).

The most successful scheme was sponsored by the Earl of Bedford in 1606 and involved cutting a watercourse from the Ouse at Earith to Denver in Norfolk. The 'Old Bedford' as this watercourse was called was followed later by a second cut the 'New Bedford'. This plan, devised by the Dutch engineer Vermuyden, aimed to cross the fens with dykes and rivers which were higher than the land and to then pump the water from the land by windmill, along the 'cuts' to the sea at King's Lynn. The two Bedford rivers with high outer banks and low inner, provided relief so that when the water level rose the resulting floodwater was deposited on the central plain: the Ouse Washes. Not only did this scheme totally change the agricultural use of the fens, but it also provided, in this relatively small area, an alternative haven for the birds displaced from the previously flooded areas.

The same principle was used on the Nene although to date its ornithological importance has remained subordinate to its better-known neighbour. Being now under management and having the advantage of being closer to the sea, this area may well prove the more exciting over the next ten to twenty years (already Black-winged Stilts have attempted to breed).

The fenland area, like the rest of the county, has one or two gravel pits, at Mepal/Block Fen and Wimblington, for example, and on its western

Figure 1. Cambridgeshire – map to show the areas of fenland and high ground.

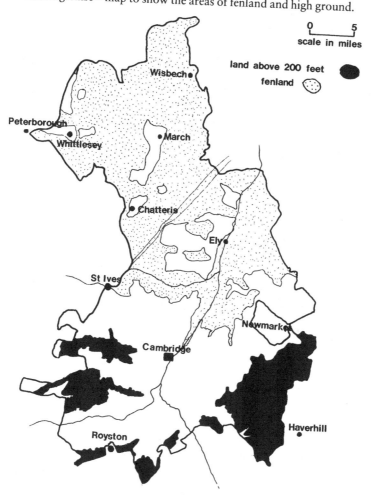

perimeter the pit at Fen Drayton is undoubtedly an important area for wildfowl in its own right.

The flatness of the fenland makes it ideal hunting ground for predators (Barn Owl, Kestrel, Short-eared Owl, and in winter the occasional Peregrine). It is also important to note that the absence of hedgerow in the fens is almost entirely due to the presence of the ditches since these form a most effective boundary between fields.

Where the fens begin or end is difficult to determine precisely but the map (Fig. 1) shows the outline of the southern limits.

A final word of explanation is required concerning the naming of parts of the water and road systems. The dykes or ditches that run along individual fields are often called simply 'drains'. In some parts the more substantial waterways, which are the size of small rivers, are called lodes, as in Wicken Lode, Burwell Lode, and Reach Lode. A further common name, drove, applies to the straight single-lane rough roads that lead to isolated communities.

CHALK

The southern half of the county is based on two differing soil types, the largest of which is the chalk escarpment that runs from Royston in the west to Newmarket in the east and covers the whole of the south-eastern part of the county. Until the Enclosure Acts, around 1800, most of this area was open grassland such as now remains around Newmarket Heath, but in time it has become good agricultural land and is now used to grow cereals, rape and sugar beet. In the chalkland, too, are to be found almost all of the ancient walkways, the Icknield Way being perhaps the best known, and the two earthworks, Fleam Dyke and Devil's Dyke, or Devil's Ditch as it is sometimes called. These were dug for the protection of the local population and are considered to be post-Roman.

The chalkland is the home of many unusual bird species and is still hedged in places, although most of the farms in the area have enlarged their fields over recent years. One of the interesting features of this part of the county are the strips of scots pine that are often used as field separators.

In this area there are many small woodlands, but their quality as ornithological sites, where it is known, is generally not high. Nor are they very great in size and the diversity of tree species is often low with ash being predominant. However, the beechwoods, of which Wandlebury is probably the finest, offer a valuable habitat to many bird species for which the county is not generally environmentally hospitable, such as the Nuthatch.

The two aspects that add to the flavour of the chalk area are based on water. First there are a number of medium-sized gravel pits that have been excavated (see Fig. 2). These provide standing water suitable for grebes, ducks, visiting waders, and many waterside passerines. Secondly, natural water is provided by the presence of springs in many small sites, Fowlmere watercress beds being the best example. This spring water which erupts throughout the year gives these sites a true wetland vegetation and has led to a number of areas on the chalkland being given the name fen, as in Dernford Fen, Quy Fen and the area called Fulbourn, Wilbraham and Teversham Fen. The woods at Sawston Hall, and at Whittlesford Middlemoor (now sadly partly ploughed up) remain damp underfoot even in summer due to springs and although small in size have a rich diversity of flora and fauna.

Over the chalkland as a whole the scrub layer proceeds rapidly where there is no grazing or cultivation to prevent it and dogwood (*Cornus sanguinea*), spindle (*Euonymous europaeus*), wild roses (*Rosa canina, R. rubiginosa*), wayfaring tree (*Viburnum lantana*) and buckthorn (*Rhamnus catharticus*) emerge, but in general the area remains open agricultural land with minimal amounts of pasture and diminishing hedge cover.

BOULDER CLAY

There are two large areas of boulder clay; one runs across the south-western part of the county from Coton to Tadlow and the other is on the eastern border from Linton to Cheveley. Their importance is that in these two areas lie woodlands of real value to the avifauna. In the west are Hayley, Buff and Hardwick Woods, Overall Grove and Knapwell Wood, and in the east Ditton Park and the Widgham Woods together with many smaller units.

The traditional method of management of these woods has been coppicing with standards which involves leaving some standard trees and coppicing most of the remainder. Commonest are standard oak and ash, with hazel coppiced beneath. The two layers give two habitat areas, with arboreal species such as tits, woodpeckers, Nuthatch, etc. catered for above, and warblers and thrushes in the scrub layer. Many of these woods have pools or streams within them adding a further vital dimension.

Apart from the woodland, however, this area like the rest of the county is mainly agricultural land interspersed with areas of human habitation.

SUBURBIA

Suburbia is a strangely neglected but important wildlife habitat which forms an important part of the southern half of the county for this is the centre of population. This heading is taken to include all areas of human habitation.

A development of this century has been the care and attention people have begun to give their gardens and although the formalisation has led to the loss of the 'wilderness garden' many of the developments have been responsible for increases in certain species. Tree-planting in general and conifer planting in particular has enhanced the suburban possibilities for nesting and shelter and there can be no doubt that the hedges have begun to assume great importance particularly where they border farmland. Nestboxes affixed to apple trees, the fallen apples, soft fruit and the regular provision of food such as nuts and seeds play a great part in providing alternatives for species whose natural habitat is shrinking away. Playing fields with their large areas of open grassland are very common around the city of Cambridge. Allotments, ponds and copses provide a very diverse habitat for those species prepared to live close to habitation. Many species can be seen in the centre of the city where the college gardens play an important role in the provision of food and shelter. Black Redstarts, for example, have a particular liking for this habitat as nearly all of the recent breeding records show.

Cambridgeshire, then, is a unique combination of habitats based on these different soil types. Its most obvious features, however, are the huge agricultural area, the lack of heavy industry and the fact that apart from the area around Cambridge City there are very few centres of population.

2
THE CHANGING STATUS OF CAMBRIDGESHIRE'S AVIFAUNA 1934-1986

As it is now more than 50 years since the publication of Lack's *Birds of Cambridgeshire* it is inevitable that many changes, both anthropogenic and topographical, have taken place. These have been influenced initially by the Second World War and latterly by the technical and agricultural revolutions that have followed in its wake.

POPULATION EXPANSION

Throughout the world the effects of population growth have both directly and indirectly been the most important factors affecting the environment and thus the avifauna, by placing great pressure on the available land for human needs such as food production, housing and recreation. For example, in my native village of Sawston, which has always been designated an expansion zone, one third of the agricultural land available in 1947 had been built upon by 1975. This is somewhat extreme, and other villages have grown less quickly, but it gives a clear indication of the sort of loss of habitat that wildlife endures in such circumstances.However, in general Cambridgeshire has retained its principally agricultural status and has not been in the forefront of the population explosion, remaining very much a rural county with any large-scale expansion limited to its principal city and some of its largest villages.

The primary effect, in Cambridgeshire, of population growth is the need for increased food production, which in this country has been coupled with a curiously illogical EEC farm policy and supported by Government development subsidies. This has led to considerable destruction of the valuable field fringes and the draining and annexing of wet meadowland and scrub in order to increase the area of land actually used for production. The resulting destruction of hedgerows, the introduction of new crops such as oilseed rape, the commercial growing of many domestic vegetables, and the loss of meadowland resulting from the change of emphasis from livestock to arable have also changed the physical nature of the county, particularly away from the fenland.

There is, however, a very positive side to the advance in technology and growth in human numbers and that is the great upsurge of interest in the environment as a whole and birds in particular. This interest has led to the voluntary organisations becoming more powerful and certainly more wealthy so that we are now blessed with several excellent reserves. The development of the Ouse and Nene Washes, Wicken Fen and Fowlmere watercress beds by carefully planned management has provided some recompense for the loss of other areas, the most notable of which was the modernisation of Cambridge sewage farm which reduced this site of national importance to one of mere local interest.

There are at the time of writing a number of important sites that remain in private ownership. This leaves their future rather insecure, and although some are sympathetically managed past experience has shown that this situation can change literally overnight.

There is, however, a most direct effect of the increasing interest and that is that more people (136) contributed to the 1984 *Cambridge Bird Club Report* than even belonged to the club in 1934 (99).

THE ENVIROMENT

The effect of the war

Following the publication of Lack's *Birds of Cambridgeshire* in 1934, the first and most vital factor that affected the county was the Second World War. Since the blockade from Europe was so effective the exhortation for self-sufficiency was used to convert 'waste' land to agricultural use, as at Burwell where areas adjoining Wicken Fen, the Adventurers' Fen of Ennion's excellent book, were drained never to return to their natural state. This undoubtedly reduced the richness of the area for wintering wildfowl and birds such as Bittern, harriers, etc. During the war years, however, the absence of some members of the community such as gamekeepers resulted in an increase in numbers of 'vermin' bird species, Sparrowhawk being a good example, so not all effects were negative. It was also around this period, if not earlier, that the old hay-meadows began to dissappear. This was a slow process linked to the gradual modernisation of farming methods, but was particularly bad for the Corncrake as noted by Norris in the *Cambridge Bird Club Report* for 1938.

Pollution

Just as a kind of stability was returning during the 'boom' post-war period with tractors replacing horses and combine harvesters taking on the harvest where reaper and binder and threshing machines had once work-ed, the horror of the effects of the organo-chlorine pesticides began to emerge particularly for those species at the head of the food chain such as Sparrowhawk and Barn Owl. Since so much of the county is an arable growing area, the extent and duration of the effects were greater than in most counties and the early 1960s proved a difficult period for most raptors. Some of the species that were affected have not recovered fully to this day.

The next environmental problem was that of Dutch Elm disease which had dual effects. Undoubtedly in a county so bereft of trees, to lose so large a proportion has proved detrimental to the avifauna and since many elms were to be found on farmland, in hedgerows or in small copses, this loss has been particularly tragic. In recompense, woodpeckers and tree feeding species benefited temporarily from both the increased food supply and the increase in available nesting holes. However, many species have been deprived of their only refuge. Rooks in particular seem to have been slow to colonise other tree species at their traditional nesting sites and have shown a steady decline up to recent times.

Once again the Kestrel has become a conspicuous feature of our countryside – a pleasant return of population after the slump in numbers during the 1960s.

The agricultural revolution

With the advance of mechanisation in the fields much has changed, for example the amount of time that a field stands fallow, when it provides opportunity for many species to forage, has been greatly reduced in recent years. The change to the environment which has caused perhaps the most dramatic effects is the destruction of large numbers of hedgerows, many of which dated back to the time of the Enclosures. This has been encouraged by the provision of Government money and the high price of the grain that can be grown on the extra land. Hedgerows have not been the only victim here, some copses and field fringes have been destroyed, wet meadows have been drained and much pastureland has gone under the plough. Cattle and sheep are no longer commonly to be seen in the county. At the time of writing, the public has become alarmed and distressed at this use of revenue and it seems likely that this policy will soon cease or even be reversed; yet not before Cambridgeshire, once again at the forefront of agricultural change, has been turned into a vast open cereal, rape and beans prairie.

Gravel Extraction

One of the most beneficial changes to have taken place in the post-war period has been the commercial exploitation of the building materials that lie beneath the surface. Clay, gravel and chalk have all been excavated to some considerable depths leaving huge voids which readily fill with water. In Cambridgeshire, while some of the chalk and clay pits are quite old, most of the gravel pits are of more recent origin, several dating from the building of the M11 motorway. One of their finest features is that they are well spread across the county (see Fig.2), and the digging of a pit near Whittlesford in south Cambridgeshire has changed the avifauna in that area to so great an extent that in two weeks recently I was able to record five different species of geese, something that must be unprecedented in that area. An additional important consideration is that most of the counties bordering Cambridgeshire now have a similar number of such pits, Huntingdonshire in particular, and the development of Grafham Water in that county cannot be ignored when seeking to explain the increase in waterbirds that has occurred. The breeding of both Cormorant and Common Tern are recent examples of the success of the pit system, and the colonisation of this country by the Little Ringed Plover is well documented.

CHANGES IN THE STATUS OF THE BIRDS

Thorpe in his lecture of 1974 to the Cambridge Bird Club stated that it was his impression that the decline of some species had begun a long time previously and that for many of the rarest the problem was that their position in this country has always been tenuous due to being on the very edge of their range.

Birds no longer breeding

With regard to the birds that have been lost as breeding species, Cambridgeshire has been no more than a reflection of a national picture; Corncrake, Nightjar, Wryneck, Red-backed Shrike and Woodlark being prime examples. For most of these species pressure, however slight, on their breeding habitat seems to have reduced their populations to untenable levels. Some species (e.g. Sparrowhawk and Barn Owl) have declined for specific reasons and now show some sign of recovery, albeit slow; others such as Quail and Stone Curlew seem to just hang on, neither increasing nor decreasing, yet their future looks rather bleak. Redstarts, having bred regularly and in some numbers in the last century, gradually declined so that by 1940 regular breeding ceased. Since then there have been spasmodic records, the last being in 1974. Whinchat also was a

Figure 2. Cambridgeshire Pit System, areas of man made gravel and mineral workings now flooded.

regular breeding species particularly on the washlands, but this species also declined from about 1966 onward and the last record was in 1974 with a possible nesting in 1981. Stonechat, which bred at the time of Lack's book, became irregular soon after and the last (possible) record was in 1961. For these species it may well be that both habitat changes and the fact that they are on the edge of their ranges have effected their demise. Throughout this century Montagu's Harrier seems to have bred spasmodically except for a period of regularity in the 1940s and 1950s, but although it had not been recorded nesting since 1959 a record in 1981 shows that it is probably premature to talk of extinction in the county.

Birds now breeding

Despite these losses there have been many gains over the period. The most spectacular example is the Collared Dove which was first recorded in Cambridgeshire in 1964 and within ten years was breeding in every village. In a more modest way Cetti's Warbler and Firecrest are two other recent colonists and although the former is likely to remain vulnerable to hostile weather conditions in winter the latter appears to be occurring more regularly year by year and could easily become a regular breeding species in the county.

The development of the various gravel pits has provided sites for Common Tern, Cormorant, Little Ringed Plover and Ringed Plover. The Ouse Washes have played a most important part in the increase of wintering wildfowl, particularly Bewick's and Whooper Swans, Wigeon and Pintail. These increasing winter numbers have led to breeding by Pintail and possibly Wigeon. Oystercatchers and Shelduck have also used the washes to breed for the first time and seem now to be well established. Wintering Hen Harriers have increased over the last ten years with good numbers roosting on both the Ouse Washes and at Wicken Fen, and a recent oversummering record; in addition Marsh Harriers have bred at least once in the county in recent years. The most spectacular returns that have been noted are Black-tailed Godwit, Ruff, Black Tern, Bearded Tit, Spotted Crake, Hobby and possibly Savi's Warbler, giving the county a certain uniqueness and showing the importance of the wetland reserves.

On the other waters grebe numbers have risen, and Canada and Greylag Geese, Tufted Duck and Pochard have all established breeding sites on the gravel pit complex. It is probably a matter of time before several other species of wildfowl breed; species with feral populations are increasingly recorded including Ruddy Duck, Egyptian, Barnacle, White-fronted and Pink-footed Geese.

Perhaps the most unlikely and exotic of all our recent colonists are the

Golden Orioles that have bred for several years in our generally cool and wet climate.

Considered as a whole the profit and loss account shows a healthy balance; however without the control and protection of our wetland reserves this picture would undoubtedly be much less favourable and unless further protection is afforded to important sites there will be little room for further advances. To hope for the return of those lost species seems futile for it is most unlikely that the conditions they prefer will ever return to Cambridgeshire.

CHANGES IN THE CAMBRIDGE BIRD CLUB

Throughout the 50 years being considered here the county avifauna has been monitored and published by the Cambridge Bird Club, which was founded in 1924 in memory of Professor Alfred Newton. More information about the present activities of the club is given at the end of the book (see Appendix).

The club has always been a great example of a successful marriage between 'town and gown'. In the early days (1930s and 1940s) the town part of the blend was founded on two people: William Farren and Maud Brindley. William Farren was by all accounts a most accomplished all-round naturalist and a taxidermist by trade who kept a shop in Cambridge which was said to be the meeting point for all naturalists in the county of any note. Thorpe, in his obituary published in the *Camb Bird Club Report* for 1953, stated that Farren was thoroughly opposed to there being separate sections of the Cambridge Natural History Society and thus was not altogether in sympathy with the club when it was first formed. However, it seems that he was not unhelpful and by the time of his death in 1952 at the age of 87 he had made a considerable contribution. Maud Brindley was clearly one of those highly committed ladies that abound in the histories of the ornithological organisations, yet unfortunately little is recorded for posterity about her work in the club since when she died, in 1941, only a short obituary was printed and promises of a more extensive version were not fulfilled.

Early university members of the club included C.W. Benson, R.A. Hinde, David Lack, N.W. Moore, Peter Scott and W.H.Thorpe, all later ornithologists; E.A.R. Ennion and J.G. Harrison, doctors by profession and ornithologists by inclination, and R.H. Adrian, A.L. Hodgkin, and R.D. Keynes, all ornithologists who became physiologists. Comprising as they do a Nobel laureate, six Fellows of the Royal Society, five professors, a Baron and two Knights they gave the club a rather high standard to maintain!

A flavour of the character of the early period of the club was given by W.H.Thorpe in his lecture 'Recollections of ornithologists and other naturalists' which was published by the Cambridge Bird Club in 1976 and contains a wealth of information on the characters of the period.

In the 1950s and 1960s the emphasis changed with the town in ascendancy and the work of the club was dominated by the efforts first of A.E. Vine and later G.M.S. Easy who between them raised both the level and volume of bird recording in the county and turned the Cambridge Bird Club Report into an annual and comprehensive systematic list rather than the previous selected observations of certain species. During this period the height of activity was at Cambridge sewage farm which had reached its most productive period; sadly a final flowering before the rapid decline as a site of importance that followed its modernisation. If William Farren and David Lack were the original Titans of the club then Graham Easy assumed that position by the late 1950s and the initials G.M.S.E. appear so much in the reports that at times other members' records seem to act only as a supplement. He has written a great deal about the county avifauna and as well as his interest in identification his published work shows his depth of knowledge and interest in species that many would consider unglamourous, such as the Rook.

The undergraduate population of the period was not without its contributors; C.J. Cadbury, P.E. Evans, Chris Mead, C.D.T. Minton and I.C.T. Nisbet are names that will be familiar to anyone who reads about birds.

In the 1970s the club underwent a change of constitution and the democratic control passed to an elected council which comprised mostly town members. In this period the interest in ringing which had been slowly building up saw the formation of the Wicken Fen Group as part of the BTO's efforts to investigate Acrocephalus warbler populations. M.J. Allen was the prime mover, supported by N.J.B.A. Branson and C.J.R. Thorne, and the aim was to concentrate resources on a single site to maximise the return for effort. To date the group has been a most successful sub-unit. Among the membership in this period were a number of those who could now be described as professional ornithologists: C.J. Bibby, N.B. Davies, A.B. Gammell, R.E. Green, D.G.C. Harper, D.R. Langslow, J.H. Marchant, J.Sorensen and T.J. Stowe.

However, the process of recording and monitoring countywide has continued to rely on Graham Easy, the present recorder Colin Kirtland and, for a while, the report editor of the 1970s and early 1980s, Tom Talbot.

REFERENCES

ENNION E.A.R. *Adventurers' Fen.* Methuen. 1942.
NORRIS C.A. Notes on the distribution and Status of the Corncrake with particular reference to Cambridgeshire. *Camb. Bird Club Report* for 1938.
THORPE W.H. The early Cambridge Bird Club: Recollections of Ornithologists and other Naturalists. Camb. Bird Club. 1976.
THORPE W.H. Obituary of William Farren. *Camb. Bird Club Report* 1953.

3
WHERE TO WATCH BIRDS IN CAMBRIDGESHIRE

W.J. Jordan

This review of good birdwatching sites presents only those areas with public access or open to members of national or local organisations. Sites such as the Ouse and Nene Washes, Wicken Fen and Hayley Wood will obviously be familiar to experienced birdwatchers, but visits to less familiar sites, particularly the wetland areas and some of the woodlands on the eastern perimeter of the county are recommended since the avifauna of many of these areas is totally unrecorded. To give an idea of the richness of these sites, over 140 species are regularly seen and over 100 breed each year. The areas are listed under the following habitat types: Wetlands, Gravel Pits and Lakes, Woodlands and Parkland, Meadows and Farmland. For each site details of location, size, access and the more unusual bird species recorded (1975-86) are given together with recent breeding (number of pairs) or wintering numbers.

WETLANDS

Cam Washes   *TL 531687-537730*

Wet meadowlands running alongside the River Cam from Upware to Dimmocks Cote
Access: There is a public footpath along the raised banks.
Size: Approx. 350 acres
Birds: Summer – breeding Redshank (17), Snipe (24), Yellow Wagtail (18), Ringed Plover, Heron (8) and Oystercatcher (1). Winter – ducks, geese and swans in small numbers.

Chippenham Fen   *TL 645695*

Woods, wet meadow and reedbed.
Access: By permit only from: Nature Conservancy Council (60 Bracondale, Norwich) local office. Closed April-June. Alternatively, a public footpath runs across the fen; leave cars at TL 652692
Size: 230 acres

Birds: Summer – Reed, Sedge and Grasshopper Warblers, Kingfisher (2), Green, Great and Lesser Spotted Woodpeckers, Nightingale (2), Nuthatch (10), Woodcock and Redshank. Wood Warbler has been reported.

Ely beet factory
TL 565808

Settling beds and lagoons.
Access: View from public road, there is no general access.
Size: Approx. 70 acres
Birds: Summer – breeding Black-headed Gull, Redshank (2) and Shelduck. Autumn – passage waders. Winter – ducks.

Fowlmere watercress beds
TL 407642

Reedbed, hawthorn scrubland, watercress beds. A large hide on stilts and a smaller by a watercress bed.
Access: An RSPB reserve with unlimited access (if privacy is preferred early morning or late evening recommended).
Facilities for wheelchairs.
Size: 66 acres
Birds: Summer – breeding Water Rail (6), Snipe, Reed, Sedge and Grasshopper Warblers, Nightingale (2) and Kingfisher (1). Autumn – passage Green Sandpiper. Winter – roosts of finches and Pied Wagtail, Siskin Water Rail and Hen Harriers (occasionally).

Fulbourn Fen (Teversham/Wilbraham Fen)
TL 520593

Reedbed, hawthorn scrub and wet meadowland.
Access: No public access to fen. Access via footpaths as follows: from bridge on the A1303 at TL 509594 along the south side, and on the north side from Frog End, Little Wilbraham, along bridleway towards the A45 interchange.
Size: Approx. 200 acres
Birds: Summer – breeding Redshank, Ringed Plover, Mute Swan and Snipe. Autumn – passage waders. Winter – Short-eared Owls, thrush roosts and feeding on haws.

Middle Fen and Mare Fen, Swavesey
TL 355696 *and* TL 366697

Wet fen/meadowland.
Access: View from several footpaths from Swavesey and along the Great Ouse. Mare Fen is a CWT reserve.

Size: Approx. 240 acres
Birds: Summer – breeding Snipe (4) and Redshank (1). Winter – owls.

Nene Washes, Whittlesey   *TL 277992*

Wet meadowland, marsh.
Access: RSPB reserve, visiting by arrangement with the warden, G.Welch,
21a East Delph, Whittlesey, nr Peterborough.
Size: 500 acres (reserve), total 1700 acres
Birds: Summer – breeding Yellow Wagtail (51), Redshank (48), Oyster-
catcher (7), Ruff (2), Snipe (250), Water Rail and Short-eared Owl (2 in
1984). Winter – Bewick's Swan (1000), Wigeon (6000), Pintail, Teal, Hen
Harrier and Short-eared Owl.

Norwood Road, March   *TL 417980*

Marsh and scrubland.
Access: CWT reserve, with a nature trail.
Size: 6.5 acres
Birds: breeding warblers.

Ouse Washes (Welches Dam)   *TL 470860*

Wet meadowland, reedbed, marsh.
Access: RSPB and CWT reserve. Open generally all day, no permit required. Visitor centre open at weekends. Ten hides, a lot of walking!

Sunday visits are recommended in the period September to January since shooting is prohibited on that day.

Size: 2000 acres (RSPB), 430 acres (CWT)

Birds: Summer – breeding Black-tailed Godwit (20), Ruff, Snipe (500), Redshank (200), Oystercatcher (8), Water Rail, Common Tern (4), Garganey, Shelduck, Heron, Kingfisher, Spotted Crake and Yellow Wagtail, among many species. Autumn – passage waders and hirundine roosts. Winter – wildfowl, particularly Bewick's and Whooper Swans, Pintail, Wigeon, Short-eared Owl, Merlin, Peregrine and Hen Harriers.

Quy Fen

TL 513628

Reedbed, pits and grassland.

Access: Along public footpaths on the south and north sides.

Size: 108 acres

Birds: Redshank, Reed and Sedge Warbler breeding.

Shepreth L-Moor

TL 385475

Scrub, slight marshland.

Access: A CWT reserve, open to members, with public footpaths traversing.

Size: 18 acres

Birds: finches, warblers, Cuckoo.

Wicken Fen

TL 563701

Reedbed, sedge fen, willow/hawthorn scrub, some woodland.

Access: National Trust reserve, generally open all day. For non-members there is an entrance fee (around £1). Two hides. A long boardwalk designed for wheelchairs was laid in 1986.

Size: 680 acres

Birds: Summer – breeding Woodcock (5), Long-eared Owl (3), Sparrowhawk, Nightingale (5), Cetti's Warbler, Bearded Tit, Reed, Sedge and Grasshopper Warblers, Yellow Wagtail and Spotted Crake, among many species. Autumn – some passage waders, huge hirundine roosts in some years with attendant Hobbies. Winter -wildfowl of many species, Hen Harriers at dusk, Woodcock, Bittern, Great Grey Shrike and Sparrowhawk.

GRAVEL PITS AND LAKES

These areas of open water are easily viewed and thus well recorded. In general Tufted Duck, Pochard and Coot are found but each site has its own speciality, notably Little Ringed Plover, passage waders and hirundine roosts. Of all the pits the finest, ornithologically, are the Fen Drayton GPs.

Bassenhally Pits, Whittlesey
TL 287986

Access: A CWT reserve, open to members, but closed on Sundays.
Size: 22 acres

Cherry Hinton Cement Pits
TL 478575

Access: By public footpaths.
Size: Approx. 15 acres
Birds: Autumn – passage waders. Winter – Great Crested and Little Grebe, Cormorant, Kingfisher and various wildfowl.

Fen Drayton Gravel Pits
TL 335687

Access: Encompassed by public bridleways.
Size: 500 acres
Birds: Summer – breeding Great Crested Grebe (2), Little Grebe (10), Shelduck (2), Little Ringed Plover (3), Redshank (2), Common Tern and Sand Martin (100). Winter – wildfowl, large numbers and many species.

Gray's Moor Pit, March
TL 414006

Access: CWT reserve, open to members.
Size: 15 acres
Birds: Great Crested Grebe, Water Rail, Sand Martin roost.

Impington Lake
TL 450620

Access: View from the A45 by-pass, lay-by on railway bridge.
Size: Approx. 15 acres
Birds: Summer – breeding Great Crested Grebe (2).
Winter – wildfowl.

Manea Pit
TL 489892

Access: CWT reserve, public access.
Size: Approx. 10 acres
Birds: Great Crested Grebe.

Mepal Gravel Pits
TL 425830

Access: View from the public road, picnic site.
Size: Approx. total 30 acres
Birds: Winter – wildfowl, often used as an overspill area by birds from the Ouse Washes.

Milton Gravel Pits
TL 480625

Access: Public access from the recreation ground.
Size: Approx. 20 acres
Birds: breeding warblers and wintering wildfowl.

Roswell Pits, Ely
TL 547805 and TL 549800

Access: CWT reserve (nature trail at second map reference)
Size: 80 acres
Birds: Summer – breeding Kingfisher and warblers. Winter – wildfowl and passage waders.

Waterbeach Gravel Pits
TL 483674

Access: Public bridleway by A10 side of the marina, limited viewing from the A10 lay-by.
Size: Many pits, some small, some large.
Birds: Summer – breeding Great Grested Grebe (1), Little Grebe (5) and Tufted Duck (5). Autumn – passage waders. Winter – wildfowl.

Whittlesford Gravel Pit
TL 464494

Access: View from public road.
Size: Approx. 15 acres
Birds: Summer – breeding Little Grebe (1), Canada Geese (3), Yellow Wagtail (3). Winter – wildfowl.

Wimblington Gravel Pit TL 435901

Access: View from public road.
Size: Approx. 10 acres
Birds: wildfowl, hirundines.

WOODLANDS

As can be seen in Chapter 1, the woodlands are based on the areas of boulder clay running in the western and eastern sections of the south of the county. Many of them are poorly documented, if at all, and a large proportion of those on the eastern edge are privately owned and accessible only along footpaths. The finest wood is undoubtedly Hayley, near Longstowe.

Buff Wood, Hatley TL 283509

Mainly elm, some oak, hornbeam and hazel. Coppiced since 1955, four ponds.
Access: Cambridge University property. The north end is open to the public. To obtain a permit for the remainder apply to the Director of the Botanic Garden, Cambridge.
Size: 39 acres
Birds: Great Spotted Woodpecker. No systematic recording done.

Ditton Park Wood, Wooditton TL 667570

Forestry Commission coniferous forest.
Access: No public access but can be viewed from public footpath.
Size: Approx. 100 acres

Eversden Wood TL 346528 and TL 343533

Mixed Woodland
Access: Private woodland traversed by public footpaths.
Size: Approx. 80 acres
Birds: Summer – breeding Great Spotted Woodpeckers and warblers.

Fordham Woods TL 633701

Damp woodland.
Access: A CWT reserve, open to members.

Size: Approx. 20 acres
Birds: Summer – Nightingale, warblers and woodpeckers.
Winter – Siskins, and titmice.

Hardwick Wood   TL 354572

Mainly oak, ash and elm.
Access: CWT reserve, open to members.
Size: 40 acres.
Birds: Great Spotted Woodpecker (2), Green Woodpecker and Nightingale (2).

Hayley Wood, Longstowe   TL 294534

Mainly oak, some ash and hazel, coppiced with several small ponds.
Access: CWT reserve, open to members or by permit.
Size: 122 acres
Birds: Summer – breeding Nightingale (4), woodpeckers, Nuthatch, Spotted Flycatcher, six titmice species, warblers, Woodcock and Cuckoo.

Knapwell Wood   TL 333607

Mixed woodland.
Access: CWT reserve, open to members.
Size: 11 acres
Birds: warblers.

Overhall Grove, Knapwell   TL 337633

Elm, oak and ash.
Access: CWT reserve. open to members.
Size: 43 acres
Birds: Tawny Owl, woodpeckers and warblers.

Papworth Wood   TL 292628

Mixed woodland.
Access: CWT reserve, open to members.
Size: 18 acres

PARKLAND, MEADOWS AND FARMLAND

In this category are a number of varied sites all of which can prove exciting and interesting. It would, however, be fair to say that they do not hold species in the concentrations of some of those sites already mentioned.

Anglesey Abbey, Lode   TL 534622

Historic home with considerable grounds both formal and informal.
Access: National Trust property, open to the public, entrance fee for non-members.
Birds: Summer — breeding finches and warblers. Winter - thrushes and finches.

Ashwell Street

An ancient pathway running from Melbourn to Ashwell through chalk farmland.

Botanic Garden, Cambridge   TL 455573

A formal garden with a small lake and a number of interesting and mature trees.
Access: Cambridge University property open to the public in general during the hours of sunlight, except on Sundays, when it is closed in the morning.
Birds: Summer — breeding finches, warblers and Spotted Flycatcher and often Little Grebe. Winter — finches, thrushes and occasionally Siskin, Kingfisher, Little Grebe and woodpeckers.(A comprehensive list of birds recorded up to 1982, written by D.G.C. Harper, is available at the office.)

Devil's Dyke

An ancient earthwork running from Reach to Stetchworth. This walk takes in some of the best fen farmland area before moving onto the old heathland of Newmarket and the wooded area at Stetchworth.
Birds: Species that can be found include all the fenland species and at Stetchworth, Nuthatch and possibly Hawfinch can be seen.

Fleam Dyke

Ancient earthwork running from Fulbourn to Balsham.

Fulbourn Meadows (Fen)   *TL 526557*

Wet, marshy.
Access: CWT reserve, open to members.
Size: 67 acres
Birds: Summer – Grasshopper Warbler (6) and Redshank (1).

Icknield Way

An ancient path running, in Cambridgeshire, eastwards to Royston through chalk farmland. Farmland species and one or two special visitors such as Dotterel in May, Quail in summer and Golden Plover in winter can be seen.

Roman Road

Another ancient track running from Cambridge to Horseheath. Farmland species, particularly Golden Plover in spring.

Soham Meadows   *TL 607721, TL 609725 and TL 610727*

Access: A CWT reserve on either side of Soham Lode. Public footpaths.
Size: Approx. 100 acres
Birds: Snipe, Great Crested Grebe.

Thriplow Meadows   *TL 437467*

Access: CWT reserve open, to members.
Size: 11 acres
Birds: Snipe, Spotted Flycatcher.

Wandlebury, Gog Magog Hills   *TL 491533*

Beech woodland and parkland.
Access: Open to the public, contributions box.
Birds: woodland species, particularly Brambling, Nightingale, woodpeckers and Nuthatch.

Wimpole Hall   *TL 343511*

An historic home.
Access: National Trust property, entrance fee. Public footpaths cross the parkland.

Birds: Nuthatch, woodpeckers, warblers. (Full list available from office at Wimpole.)

THE SYSTEMATIC LIST

INTRODUCTION

The list presents a summary of or, in certain cases of rare species, all the published information on every species for which there has been an acceptable county record since scientific recording began and the sequence and nomenclature are those recommended by K.H.Voous in his *List of Recent Holarctic Species* (1977).

For each bird this information is presented in three main sections:

Pre 1934

This section is drawn mainly from Lack's *The Birds of Cambridgeshire* (1934), augmented by the writings of Leonard Jenyns together with A.H. Evans' list of avifauna in Marr and Shipley (1904). This information is included since these sources are no longer readily available. The sources are given in square brackets after each record, although many records are to be found in more than one of the above.

1934-1969

In this period the information is based on the records published by the Cambridge Bird Club in their annual reports and the various articles and papers that augment them including a further list by A.H. Evans in the *Victoria County History of Cambridgeshire* (1938).

Present Status (1970-1987), Breeding Status and Ringing Results, (where relevant)

This section, which is preceded by a short status summary, is also drawn mainly from the reports of the Cambridge Bird Club augmented by information in the scientific literature resulting from individually, or nationally, organised investigations. In the case of certain species, particularly those in the latter part of the list, personal experience and information provided by Graham Easy has been included. In an attempt to produce up-to-date summaries some information is included which is unpublished at the time of writing; furthermore, some of the records of rarities have not been processed by the British Birds Rarities committee.

In some instances, for example Heron, where status has changed very little, the sections are combined to avoid repetition. References in which the reader will find further information are included at the conclusion of the relevant species to avoid the tiresome neccesity of continually turning to the back of the book.

Certain terms have been used in a definitive manner:

Extremely rare: less than five records in all.

Very rare: less than ten records in all.

Rare: between ten and thirty records in all.

Most species with fewer than twenty records have them listed; however, in some circumstances, such as rapidly increasing birds, this is not the case.

Other terms such as common and abundant do not require explanation except that their use is rather subjective and the word uncommon has been used to imply rarity in birds that do not fit the above definition of rare.

An attempt has been made to differentiate between species which are recorded throughout the year, e.g. Lesser Black-backed Gull, and those that are sedentary and resident, e.g. Blue Tit.

Where ringed birds are discussed two phrases may require explanation. The term 'recovery' applies to a bird that having been ringed was found at another place and was not subsequently released; these are usually dead birds picked up by members of the public. The term 'control' applies to a bird that is caught, usually by another ringer, at a site other than its ringing location and released.

Earliest arrival and latest departure dates of migratory species are included, although with some migrants, e.g. Blackcap, Chiffchaff, etc. they have been omitted since they are somewhat difficult to determine due to the complication of overwintering birds.

Throughout the list, with respect to those records which refer to the Ouse Washes, every effort has been made to restrict them to those from the Cambridgeshire section; however, most of the wildfowl and wader counts which are carried out by the RSPB and Wildfowl Trust reserve staff measure the whole area including the part that is in Norfolk.

With regard to species recorded as 'escapes' past history has shown that it is quite possible that some of these may establish themselves in the wild and at some future date be added to the British List. However, since recording of these birds is very variable, according to the observer involved, I have not included them here.

Also it is a matter of great regret that over the years for various reasons some significant records have not been submitted to the county recorder and are therefore not included either.

Finally, the reader should bear in mind that there is an unfortunate natural bias in the observations since most of the population of the county is centred on and around the city of Cambridge and our coverage of the fenland and northern areas has been largely the work of one or two dedicated observers.

Those references which are used throughout the list are:

EVANS A.H. The birds of Cambridgeshire. In Marr J.E. and Shipley A.E. *Handbook to the Natural History of Cambridgeshire.*Cambridge University Press. 1904.

EVANS A.H Birds, zoology of Cambridgeshire. In Imms A.O. *Victoria County History of Cambridgeshire and Isle of Ely* Vol.1. 1938.

JENYNS L. *Towards a fauna cantabriensis.* MS deposited in the Cambridge University Zoology Library. 1869.

LACK D.L. *The Birds of Cambridgeshire.* The Cambridge Bird Club. 1934.

SHARROCK J.T.R. *The Atlas of Breeding Birds in Britain and Ireland.* T & A.D. Poyser. 1976.

THORNE C.J.R. and BENNETT T.J. *The Birds of Wicken Fen.* The Wicken Fen Group. 1982.

SEABIRDS

In a landlocked county all marine birds are found only when conditions on the coast are sufficiently severe to drive them inland. This occurs during periods of depression over the North Sea and is usually associated with exceptionally strong winds. It can be seen from the records that this is not a common occurrence. Birds that remain for any length of time are usually injured or victims of oiling. Estuarine birds are reported more commonly, almost always on passage or during the winter.

Red-Throated Diver   *Gavia stellata*

Pre 1934

Jenyns noted that it was occasionally killed in the fens. Evans merely described it as a rare winter visitor. Lack quoted four records as below:

One shot on Whittlesey Washes (Nene) in 1882.

One shot on Whittlesey Washes in 1882 (this may be the same record).

One at Ely January 1907 (reported by Farren, he told Lack that he had others).

One dead on the Gogs in December 1927 (Prof. G.I. Taylor).

There was also an undated specimen in Wisbech Museum from Guyhirn reported by Evans.

1934-1969

Records were exceptionally scarce until the 1950s, there being only two, both in 1947, in the 24-year period between 1927 and 1951. For the rest of this period there was at least one record in most years, of which over 70% were on, or around the area of, the Ouse Washes.

Present Status

An irregular winter visitor, usually either storm-driven or sick/injured birds.

At the time of writing there have been 46 records of 52 birds of this species. Most occur in the three-month period January-March and it is rare for more than a single bird to be involved. Sites preferred are: the large open-water areas (the Washes), the suitable pits nearby, and the main rivers (Ouse and Nene). A number of records refer to oiled birds which often remain on site for many days; others occur after a period of hostile weather at sea and, sadly, many records are of birds found dead. Genuine passage through the county, if it occurs at all, must be considered unusual.

After a period without records during the 1970s there has been a return to a near annual pattern.

Black Throated Diver   *Gavia arctica*

A very rare vagrant.

Seven records, all listed below.

1 One killed on Barton Pond in January 1850. [Lack]
   [One killed at Chesterford (Essex) in 1850 was reported in Lack although it is outside the county.]
2 One shot near Barton in November 1912 was sent to Farren. [Lack]
3 One on the Ouse Washes, Hundred Foot River, on 6 and 11 February 1955.
4 One at Welches Dam (Ouse Washes), 12 February 1956.
5 One on the Ouse Washes, 5 March 1966.
6 One flying north-east over the Ouse Washes on 7 January 1968.
7 One on the Ouse Washes, 7 January 1972.

Great Northern Diver   *Gavia immer*

A rare vagrant.

Eleven records to date, all listed below:

1 A young bird, shot at Burwell in November 1843. [Jenyns, Lack]
   (Evans is rather sceptical of this and a Black-throated Diver record
   which he describes as 'supposed'.)
2 One on 'the Washes', presumably the Ouse, in November 1927.
3 One at Kirtling flew into the county from Suffolk on 8 November
   1952, was caught, ringed and released unharmed at Abberton
   Reservoir, Essex.
4 One on the Ouse Washes near Welney on 5 February 1961.
5 One at Waterbeach GP on 20 November 1969.
6 One on the Ouse Washes at Purl's Bridge on 23 February 1979.
7 One at Mepal GP, 29-31 December 1979.
8 One on the Ouse Washes in mid November 1980.
9 A first-winter bird on Fen Drayton GP, 30 January 1985.
10 One at Whittlesey West Pit, 16 March 1985.
11 An adult at Fen Drayton GP, 2-12 February 1986.

Little Grebe   *Tachybaptus ruficollis*

Pre 1934

Evans described this species as 'extremely uncommon' but by the time
Lack wrote it had increased, and he reported that it was to be found
throughout the year, with parties of up to 20 on the washes.

1934-1969

Records in the middle period indicate a small breeding population at the
sites that were suitable at the time. There was also, even in the 1940s, a
tendency for the birds to gather in winter on the section of the River Cam
leading north from Cambridge; for example, 30 were seen between Cambridge and Waterbeach in 1941. This, it was suggested, was a gathering of
the breeding population. By 1947 up to ten had been recorded on the
Ouse Washes in winter and throughout the 1950s and 1960s this position
changed very little.

Present Status
Resident.

The most widespread of all the grebes, this species is found on most
suitable gravel pits and river systems. Some congregations occur, particularly in hard winters when pits are frozen; for example 31 and 37 at
Bottisham Lock on the Cam in the years 1976 and 1982 respectively.

Breeding Status

Numbers vary, but in the last ten years or so there have been between 13 and 33 pairs recorded at between six and twenty sites. These represent minima and the actual number could be much higher. It has certainly increased since Lack's time but a recent decline in population since the early 1970s suggests that all is not well. The (post-Lack) improved status is in line with the development of clay and gravel pits, but a large number of suitable looking territories, such as the vast systems of dykes in the fenland, hold a very poor population.

Great Crested Grebe   *Podiceps cristatus*

Pre 1934

Before the draining of the fens it was apparently common but by the time Jenyns was writing it had become scarce. Evans described its status simply; 'occurs but rarely'. Lack, in a lengthy summary, pointed out that it was a regular winter visitor to the River Ouse and the Ouse Washes with

Figure 3. Great crested Grebe – number of breeding pairs in Cambridgeshire 1963-84.

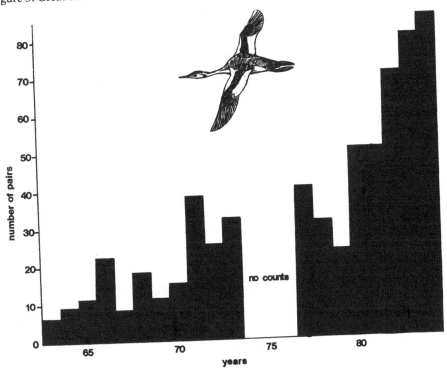

usually single birds involved. He also stated that Farren received a number of specimens and that one such c.1920 was said to be one of a pair breeding at Chesterton Ballast pits. In 1932 a nest was found there but the eggs were later destroyed. Vine (1961) suggested that breeding (or attempted breeding) took place at this site from 1912 until 1932.

1934-1969

In 1934 there was a report of breeding at a pit outside Cambridge 'where it has bred for several years'. Secrecy surrounded many of these early colonies throughout the 1930s and thus there were pairs at suitable sites without breeding being publicised. However, in 1940 a pair was reported to have bred on 'Mill Road pond'. Subsequently breeding was rather irregular until 1948 since when it has been annual. Up to the mid 1960s there was a more stable population of 4-10 pairs. Winter gatherings were also noted; on the Ouse Washes up to six in 1947, up to ten in the spring of 1951 and in the autumn in the area around the Cambridge sewage farm. The 1965 national census (Prestt and Mills 1966) revealed a total of 14 pairs on eight waters and wintering numbers had by 1967 reached a maximum as high as 25 on the Ouse Washes.

Present Status

A common resident on areas of open water, and an autumn and winter visitor.

Winter concentrations occur on both the Ouse and Nene Washes with maxima of 19 (1981) and 27 (1984) respectively. Recently numbers have built up on the gravel pits with maxima as follows: Fen Drayton 30 (1984), Mepal 20 (1978), and at Roswell Pits 12 (1979). In addition it is recorded, usually as a visitor during milder conditions, at many smaller pits such as Hauxton, Impington, Shepreth, Whittlesford, etc.

Breeding Status

A dramatic increase in breeding numbers has taken place recently (Fig. 3). From 10-20 pairs in the late 1960s numbers increased to around 80 pairs in 1983 following the colonisation of fenland dykes and riverside sites. Easy (1984) has analysed this increase and concluded that the expansion has taken place along the waterways of the county probably as a result of birds attracted to the Ouse Washes to breed being displaced to adjacent river systems as floodwater subsided.

ANDREW D.G. The breeding of the Great Crested Grebe (*Podiceps cristatus*) on Milton GP 1948. *Camb. Bird Club Report* 22. 1948.

EASY G.M.S. Great Crested Grebes in Cambridgeshire. *Camb. Bird Club Report* 58. 1984.

PRESTT Ian and MILLS D.H. A census of the Great Crested Grebe in Britain 1965. *Bird Study* 13. 1966.

VINE A.E. Breeding status of three diving species in the last fifty years. *Camb. Bird Club Report* 35. 1961.

Red-Necked Grebe   *Podiceps grisegena*

Pre 1934

Jenyns noted three occurrences; at Wisbech in 1827, at Bottisham in 1829 and at Stretham in 1832. Evans described this species as 'still rarer than the Great Crested', and noted birds from Cottenham and Cherry Hinton around 1900. Lack added a further record of a bird that had been shot and taken to Farren from Arrington in December 1904.

1934-1969

After the record of 1904 the next was not until 1939 when three were seen on the Ouse Washes near Welney; and there was a probable record from the Ouse Washes in February 1941. Thirteen years passed before the next

Figure 4. Red-necked Grebe – monthly distribution of all records.

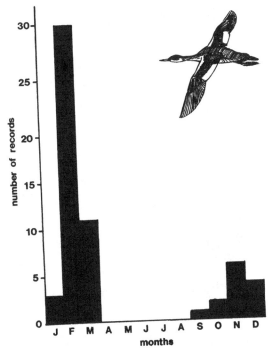

at Eldernell in December 1954. There were two more occurrences in 1955, one of which involved a bird which visited a bird-table in Great Shelford, was caught and cared for, but subsequently died. Three records in the period February/March in 1956 were followed by a gap to January 1963 when a bird found near Cottenham was 'put out of its misery'. One record in each of the years 1967 and 1968 brought the total at the end of this period to 17 in 140 years.

Present Status

A uncommon visitor, usually in winter.

In the years between 1970 and 1986 there have been 29 records, not annual but usually with two or three in the same year. There was an unprecedented number of 19 in 1979 when there was a massive influx following blizzards in mid February, and although some records may refer to the same bird it seems likely that around 14 individuals were involved. The arrival seems to have been between the 15 and 18 February. Most birds remained for four or five days, presumably to recover; the longest duration of stay was on the Ouse Washes where up to five birds were present from 15 February to 17 April. Most records fall within November/December or February/March and none have been reported before September or after April.

Slavonian Grebe   *Podiceps auritus*

Pre 1934

Evans stated that it 'occurs at intervals', while Lack gave five records, four of which were post 1900, the other being recorded by Jenyns at Bottisham in the nineteenth century.

Figure 5. Slavonian Grebe – monthly distribution of all records.

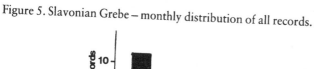
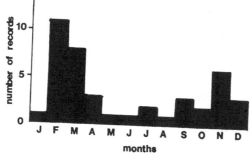

1934-1969

There were a further 12 reports in this period and almost all of them (9) were seen in the period February to April. The favoured sites were understandably on or near the Ouse or Nene Washes. The only exceptional record was of a single bird which stayed on the River Nene from February to October in 1956.

Present Status

An uncommon visitor, mostly in winter.

A further 17 records since 1970 has brought the total to 34. Almost all the records are of single birds and there has never been more than two birds together. In the recent past most have been on the various gravel pits particularly Fen Drayton. Duration of stay is usually very short (1-2 days) but can be as long as two weeks, particularly in the autumn. The presence of an adult and a juvenile at Fen Drayton on 29 July 1973, together with the bird on the Nene (see above) are the only records outside the period 24 September-15 April the overall monthly distribution appears in Figure 5.

Black-Necked Grebe   *Podiceps nigricollis*

Pre 1934

Evans stated, as for the Slavonian, that this species 'occurs at intervals'. Lack gave six records, five of which were from the nineteenth century. One was of two birds, the remainder of singles, most of whose destinies were determined by the shotgun. Farren saw one in breeding plumage on the Cam near Burwell Fen in April 1920 which may be significant in view of the later record (see below).

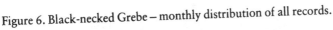

Figure 6. Black-necked Grebe – monthly distribution of all records.

1934-1969

There were 23 records in this period, mostly of single birds but four of two and three of three. The most exceptional concerns three birds at Burwell Fen in full breeding plumage in 1937, of which two remained to nest successfully providing the county's only breeding record to date. In the middle part of this period the importance of Cambridge sewage farm and nearby pits for records of this species emerges clearly when consideration is given to the number of sightings there per decade which was as follows: 1930s (1), 1940s (4 of 5 birds), 1950s (6 of 8 birds) and 1960s (1). It seems clear that the demise of the farm affected the records, the last such sighting being in 1961. There was also a strong seasonal bias towards autumn at that time with 10/23 records in the period 28 July-1 September.

Present Status

An uncommon and irregular visitor, by no means annual.
There have been eight records since 1970. Almost all were single birds at different sites, in different months (two occurrences in February), in different years, and most stayed for only one or two days. In the 26 years since the last Cambridge sewage farm report there have been only 12 records. The monthly distribution of all records is shown in Figure 6, and in contrast to the Slavonian which seems to be mainly a winter visitor, Black-necked Grebes are predominantly seen in autumn.

Black-necked Grebe raised two young at Adventurers' Fen in 1937.

Black-Browed Albatross *Diomedea melanophris*

An extremely rare vagrant.
One record only.
One was caught, exhausted, near Linton on 9 July 1897 [Lack]. This was the first British record for this species.

Fulmar *Fulmarus glacialis*

A rare vagrant.
Eleven records, all in the period March to 28 June listed below:

1 One picked up in Cambridgeshire after a gale in March 1917. [Lack]
2 One found 'long dead' at Cambridge sewage farm, 28 June 1955.
3 A dark-phase bird, found dead on the Ouse Washes on 11 March 1962, was part of a massive 'wreck' of 849 birds on the eastern side of the country which was also noted in the Netherlands and Sweden (Pashby and Cudworth 1969).
4 One flying ENE over Chatteris, 22 April 1968.
5 One found injured at Impington, 27 March 1971.
6 One flying from the north over Isleham, 2 April 1975.
7 One found dead on the Ouse Washes, 29 April 1975.
8 One found dead on the Ouse Washes, 18 April 1977.
9 One flying south-west at Purl's Bridge (Ouse Washes), 26 April 1981 was part of an eastern inland invasion and 'wreck' following gales (Nightingale and Sharrock 1982).
10 One in Trumpington, 17 April 1983.
11 One at Welches Dam (Ouse Washes), 31 May 1983.

NIGHTINGALE Barry and SHARROCK J.T.R. Seabirds inland in Britain in April 1981. Brit. Birds 75. 1982.
PASHBY B.S. and CUDWORTH J. The Fulmar wreck of 1962. Brit. Birds 62. 1969.

Manx Shearwater *Puffinus puffinus*

A rare vagrant.
Eleven records to date, all part of September dispersal from breeding colonies, are listed below:

1 A specimen sent to Farren from 'near Royston' in September 1906. [Lack]
2 One found dead at Longstanton in September 1936.

3 A dead first-year female was picked up six miles west of Cambridge on the Bedford road in the latter half of September 1947.

4 One picked up after flying into overhead cables at Wicken Fen on 25 September 1964 had been ringed as a pullus at Skokholm nine days earlier.

5 One found at Friday Bridge on 6 September 1979 was taken first to Welches Dam and then on to Minsmere where it was released the following day. The unusual feature of this record is that at the time the weather was perfect with clear skies and high pressure prevailing.

6 One found dead at Milton Road lay-by, 10 September 1980.

7 One picked up alive at Littleport, 5 September 1983.

8 One picked up alive near Three Holes, 12 September 1983. This and the previous bird were later released on the east coast.

9 One released near Burwell in September 1985 having arrived exhausted the previous evening.

10 One, exhausted, at March on 6 September 1986 was later released at Shingle Street, Suffolk.

11 An immature, found exhausted at Swavesey 4 September 1987 was cared for overnight and released at Cley, Norfolk, the following day.

Storm Petrel   *Hydrobates pelagicus*

A very rare vagrant.

Seven records, listed below:

1 One in Cambridge, spring 1828. [Jenyns, Lack]

2 One in Saffron Walden Museum from Whittlesford 1836. [Lack]

3 One at Newmarket in November 1885. [Lack]

4 One at Bottisham in November 1885. [Lack]

5 One at Reach, 2 December 1929. [Lack]

6 One found dead at Steeple Morden, 20 November 1957.

7 One killed by a cat at Elsworth on 9 July 1979.

Leach's Petrel   *Oceanodroma leucorrhoa*

A very rare vagrant.

Nine records to date listed below:

1 One at Bassingbourn in the winter of 1831/32. [Jenyns, Lack].

The following were over the period late October-early November 1952 when there was a national 'wreck' (Boyd 1954) with 6000-7000 found right across the country following a severe gale in the Atlantic.

2 One found dead at Cambridge sewage farm 31 October 1952.

3 One picked up alive at Papworth Everard, 31 October 1952.

4 One found dead in Cambridge, 1 November 1952.

5 One picked up alive at Shepreth, 1 November 1952.

6 One found dead in Cambridge, 2 November 1952.

7 One found dead in Cambridge, 3 November 1952.

8 One seen swimming, exhausted, at Fen Drayton GP on 13 November 1977 had departed by the following morning.

9 One, storm-driven, flying erratically over farmland at Whittlesford at 4 p.m. on 2 October 1978.

BOYD H. The wreck of Leach's Petrels in the autumn of 1952. *Brit. Birds* 47. 1954.

Gannet   *Sula bassana*

Pre 1934

Evans stated that it 'strays now and then into the county' and together with Lack provided details of nine records of ten individuals, most from the early part of the nineteenth century.

1934-1969

There were nine further records in this middle period, many of them of exhausted, gale-driven birds such as the one caught in a chicken run in 1952, or the one found walking along a road north of Littleport in 1964. (This was taken into care but subsequently died.) As might be expected there is a strong seasonal bias towards the autumn, and all records were of single birds.

Present Status

An uncommon vagrant.

In the period 1970-80 there were a further six records bringing the total at that time to 24. On 27 April 1981 there was a 'wreck' across eastern England (Nightingale and Sharrock 1982) with at least eight individuals recorded in Cambridgeshire: one at Cambridge, one flying north at Chippenham at 8.23 a.m., an immature at Bar Hill at 8.30 a.m., two adults and two immatures along the A604 at 8.48 a.m. Over the following days dead or dying birds were reported from Littleport, Over, near Welney and Whittlesey. There has been a single record since of an immature over the Ouse Washes in September 1984 making a total to date of just over 30 records.

NIGHTINGALE Barry and SHARROCK J.T.R. Seabirds inland in Britain in April 1981.
   Brit. Birds 75. 1982.

Cormorant   *Phalacrocorax carbo*

Pre 1934

Jenyns reported one on the top of King's College Chapel in 1825 and
another at Overcote in October 1827; this species was still nesting in
Suffolk at that time. Evans and Lack listed a further ten occurrences in the

By 1980 this isolated tree at Fen Drayton Pit had become a major roost site for Cormorant.

autumn-winter periods in the years from 1893 including one seen on Ely cathedral.

1934-1969

In the period up to 1950 there were only six records most of which were at Cambridge sewage farm.From 1950 to 1960 there was a pattern of annual records, usually only two or three and predominantly single birds either at Cambridge sewage farm or on the Ouse Washes. In the 1960s the pattern remained but the number of records increased to about seven or eight per annum and in 1965 a westward movement was observed in late March with flocks of 31 at Wicken Fen and 9 over Fowlmere watercress beds. In 1966 came the first record of attempted roosting on the Ouse Washes.

Present Status

A resident at Fen Drayton GP and on the Ouse Washes elsewhere an unusual visitor.

A complete change took place in the status of this species in the early 1970s so that from being an occasional seasonal visitor birds began to spend the whole winter on the Ouse Washes and at Fen Drayton GP. Roosting developed during this time, although the site on the Ouse Washes was initially just over the county boundary at Welney, Norfolk, and from small numbers rapidly built up to a maximum of 67 in 1977, while at Fen Drayton the roosting tree has regularly attracted 90-100 during the 1980s. Because of the numbers present in winter many birds stray to the nearby areas of open water being found occasionally at places such as Ely beet factory, Nene Washes and Wicken Fen.

Breeding Status

A pair nested at Pymore railway bridge on the Ouse Washes in 1983, the first record for the county. Since they nested in a tree it was considered possible that these birds were of Continental origin. Nest building was begun at Fen Drayton GP in April 1985 but abandoned within a few days.

Ringing Results

There have been no ringing recoveries to date.

Shag *Phalacrocorax aristotelis*

Pre 1934

Evans stated that it 'strays now and then into the county'. Lack quoted six records, only one of which was from the nineteenth century, and all between the months of October and March.

1934-1969

Up to 1953 records remained unusual. In 1937 several were seen at Mepal, in the winter of 1936/37 one was seen in Cambridge and there was a further record there in 1948. After 1953, however, there were records in most years. In 1958 there was an unprecedented invasion which began in January and led to some birds staying in the area until July. Three birds found dead during this period had been ringed the previous summer on the north-eastern coast of Britain (see below) thereby showing a uniform movement SSE. This was also true of a bird found at Wisbech in 1954. The only other recovery is of a bird from Lundy (Devon) which was found at Littleport. After a period of one or two records per annum there was another invasion in March 1962 when as many as 50 birds were present at one time; most hung around for quite a while, predominantly in the Ely-Littleport area. After 1962 the former pattern was resumed.

Present Status

An irregular visitor, usually storm-driven or ailing birds.
Still an irregular visitor, although it occurs in most winters.Single birds are found on various waterways, with the Cam being particularly favoured together with the Ouse Washes. Most birds appear to be immatures and stay for a while. An exceptional bird spent most of 1980 and the first part of 1981 on the Cam around Cambridge and was regularly seen along the college 'Backs'.

Ringing Results

There have been several recoveries in the county, usually of sick, injured or dead birds. The sources of these visitors are the breeding colonies around the British Isles, particularly the Isle of May from which two were found in 1958 and two more in 1984. There are recoveries also from Bass Rock (1958), Farne Islands (1958) and Lundy (1957).

BITTERNS, HERONS AND STORKS

Bittern   *Botaurus stellaris*

Pre 1934

Lack noted that this species bred in the county in the nineteenth century, and Evans stated that a nest was found near the River Cam as late as

1821. Many bones have been discovered in peat deposits around the fenland part of the county. Only seven records were published between 1850 and the turn of the century but by the time Lack wrote this species was more regularly observed in winter.

1934-1969

In the early part of this period individuals were commonly seen throughout the winter in the Wicken Fen/Burwell area. In 1936 at least one bird summered there and in 1937 a pair bred. Following this there were occasional summer records through to 1948 after when it reverted to being an autumn and winter visitor. In 1959 a bird was heard booming at Wicken Fen in early spring. One unusual observation was of a bird flying over King's College on 8 February 1960 but in periods of severe cold weather this species is sometimes reported away from its normal wetland habitat.

Present Status

A regular winter visitor at one or two favoured sites.

The pattern established in the 1950s and 1960s continues with regular winter sightings of one or at most two birds at the usual sites: Wicken Fen, Ouse and Nene Washes, and to a lesser extent Fowlmere, Fulbourn Fen and Waterbeach GP. Records in the part of the county away from fenland are extremely unusual and usually associated with extreme weather conditions.

Ringing Results

A bird shot at Manea on Christmas Eve 1963 had been ringed in the south of Holland as a pullus six months earlier.

Little Bittern   *Ixobrychus minutus*

An extremely rare vagrant.

Three records, listed below:

1 One, shot at Ely in 1848, is now in the Cambridge University Zoology Museum. This bird was accompanied by a female which was also shot. [Evans, Lack]
2 One labelled 'Waterbeach Fen 1858' is in the Saffron Walden Museum. [Lack]
3 A female, freshly dead, picked up under wires at Overcote 7 June 1981.

Night Heron   *Nycticorax nycticorax*

A very rare vagrant.

Nine records, five between 1980 and 1987, all listed below:
1 One near Wisbech in June 1849. [Lack]
2 One at Milton Fen, 19 April 1952.
3 One at Ely, early on 3 May 1970.
4 One on the Ouse Washes near Welches Dam, 8 June 1978.
5 A sub-adult flushed from a small pond near Wicken, 28 May 1980.
6 One at Welches Dam (Ouse Washes), 28-30 June 1983.
7 One on the Ouse Washes, 15 June 1984.
8 An immature on the Ouse Washes, 13-15 June, and in July 1987.
9 A second-summer bird on the Ouse Washes, 27 July 1987.

Squacco Heron   *Ardeola ralloides*

An extremely rare vagrant.

Three records, all listed below:

1 Jenyns noted one taken in Cambridgeshire between 1820 and 1825 (*Manual of British Vertebrate Animals* p.189) but he omitted this record from his manuscript in the University Zoology Library. Yarrell (1st Edition 1843 II p.466) referred to it. [Lack]

It seems probable that Little Bitterns have nested in Cambridgeshire.

2 One stayed on Chesterton Fen from 22 May until 3 June 1954, was seen regularly and was obviously the bird reported on 20 May over Fen Ditton.

3 One, in full breeding plumage, on the Ouse Washes, 12 June 1970.

Little Egret   *Egretta garzetta*

An extremely rare vagrant.

Two records, listed below:

1 One shot at Whittlesey about 1850. (Lilford Birds of Northamptonshire II p.128. 1895). [Lack]

2 A remarkable record of one in the grounds of the Leys School, Cambridge, 18 May 1976.

Great White Egret   *Egretta alba*

An extremely rare vagrant

One record only:

One was shot on Thorney Fen in May or June 1849. [Lack]

Heron   *Ardea cinerea*

Pre 1934

Both Evans and Lack concentrated on breeding numbers which were probably in the region of 50 pairs. Curiously Nicholson reported none in the census results of 1928, clearly the sites at Mepal, Guyhirn and Botisham were in occupation at that time although other previously known sites (Chippenham, Chatteris and Littleport) had by then been deserted. Lack finished by saying that the Heron was well distributed over the county in winter when it was more common than in summer.

1934-Present Status

A common resident.

The fluctuations in population levels are best reflected in the breeding numbers and these are both presented graphically (Fig. 7) and discussed below. Throughout this period it is fair to say that the Heron has been, and still is, a common bird right across the county, using all the river systems and areas of open water to feed. It is undoubtedly true that the atrocious weather in the winter of 1962/63 took a very heavy toll and that this was exacerbated by the effects of the toxic chemical poisoning that occurred around this time.

Breeding Status

The number of pairs at each of the heronries in the county has been counted for well over 30 years by A.E.Vine and it is thus possible to present a very consise account of the changes (see Fig. 7). In 1932 a count of nests produced a figure of 42 from three sites: Mepal, Guyhirn (which is probably the oldest site in present use) and Bottisham. By 1947 the tally was 32 and by 1950 it was 83. It remained at around this level until the slump in the early 1960s. By the early 1970s numbers had begun to climb again but while in 1977 there were 108 at five sites, Vine noted that other nearby colonies (in Norfolk) were declining. At the present time there are five principal sites at Guyhirn, Mepal, Pymore, Eldernell and Quy. Those no longer used are at Chettisham (last recorded nest in 1976) and at Bottisham (last recorded nest in 1961). In numbers the most important are Guyhirn (around 20 nests), Mepal (recent average 30-40) and Quy (around 30).

Ringing Results

Throughout the 1930s there was a programme of ringing nestlings at the

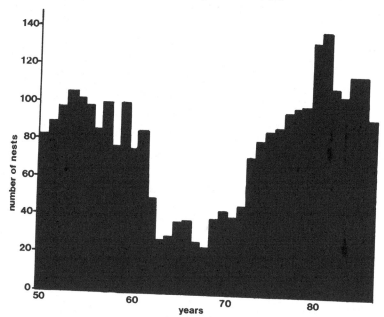

Figure 7. Heron – number of nests in Cambridgeshire 1950-84.

various sites in the county. Sadly, all the recoveries were within the first year of life and many were within a few months of fledging. The British recoveries are displayed on the map (Fig.8). Two birds crossed the channel, both from the Mepal heronry; one to Belgium and one to France. One bird found in Cambridgeshire in May 1935 had been ringed in Jutland, Denmark, in May 1934.

NICHOLSON E.M. Report on the British Birds Census of heronries 1928. *Brit. Birds* 22. 1928-29.

Purple Heron   *Ardea purpurea*

A rare vagant.

Sixteen records, all listed below:

1/2 Two obtained near Ely in the winter of 1826/27 are to be found in the Cambridge University Zoology Museum. [Lack]

3 One in Wisbech Museum was also shot near Ely and was purchased in 1845. [Lack]

4 A fourth specimen is in Saffron Walden Museum labelled 'Cambridgeshire'. [Lack]

5 A stuffed specimen was found in the possession of a Burwell farmer (George Mason) who bought it from the man who shot it (whose

Figure 8. Heron – British ringing recoveries of birds ringed in Cambridgeshire

name is not recorded) at Wicken Fen one summer between 1895 and 1900.

6 One close to the Cam in the area of Cambridge sewage farm, 16 August 1947.

7 One at Fulbourn Fen, 4-5 April 1959.

8 One on the boundary with Norfolk by the River Delph, 5 June 1967.

9 One on the Ouse Washes, 18-31 August 1978.

10 One on the Ouse Washes between Sutton and Mepal, 15 April 1979.

11 One on the RSPB part of the Ouse Washes reserve, 23 May 1982.

12 A first-year bird near Purl's Bridge, Ouse Washes, 16-27 June 1983.

13 One flew north-east past Purl's Bridge across the Ouse Washes towards Pymore, 3 June 1984.

14 One on the Ouse Washes, 28 April 1985, flew up from a ditch to be mobbed by Lapwings before flying off.

15 One on the Nene Washes, 7-11 September 1986.

16 One on the Ouse Washes, 24 June-26 July 1987 spent most of its time in the Norfolk section but made occasional visits to the Cambridgeshire part.

Glossy Ibis   *Plegadis falcinellus*

An extremely rare vagrant.

Two records, listed below:

1 One in Holywell, October 1909. [Lack]

2 One at Balsham, October 1912. [Lack]

Spoonbill   *Platalea leucorodia*

A rare vagrant.

At least thirteen records, all listed below:

1/2/3 Three birds were said to have been killed in spring 1845. [Jenyns, Lack]

4 One seen for several days at Ely, from 9 July 1877. [Lack]

5 One on Reach Fen, April 1917. [Lack]

[One was said to have visited Peterborough sewage farm in 1945 but this observation remains unconfirmed and the observer has not been traced.]

6 One on the Ouse Washes on 23 April and the same, or another, from 11 June to 2 July 1967.

7 One on the Ouse Washes in April 1974.

8 One on the Ouse Washes during April and May 1980.
9 Four moved from Norfolk to Cambridgeshire along the Ouse Washes, 24-25 May 1981.
10 Two on the Ouse Washes, 20-23 May 1982.
11 One on the Nene Washes, 17 June 1986.
12 An immature on the Ouse Washes, 5-6 June 1987.
13 One at Fen Drayton GP, 25 August 1987.

SWANS, GEESE AND DUCKS

Over the period 1934-1986 these species have generally increased, particularly in the last 20 years. Much of this is due to the purchase and subsequent protection of two remarkable sites. The development of the Ouse and Nene Washes as reserves has seen a dramatic increase in the wintering numbers of both swans and ducks in the county. Both Bewick's and Whooper Swans are now often seen around the county in small numbers flighting from these two areas, and the huge flocks of Wigeon, together with smaller numbers of Teal and Mallard that feed on the Washes themselves, give a visitor a very clear idea of how the fens must have looked prior to draining. In the 1983/84 winter the Ouse Washes held more wildfowl than any other site in Great Britain (43 217 in February) including three-fifths of the Bewick's Swans and the only significant numbers of Whoopers in lowland Britain (Salmon and Moser 1984). It was the most important site for Mallard and the second most important for Wigeon. The Nene Washes, although still developing as a reserve, are already third in the list of importance for Bewick's Swan and fourth for Pintail. Since these two areas are so close, geographically, there is of course considerable interchange of birds and many wildfowl are to be

During the winters of the 1980s Bewick's Swan flocks have become regular 'landmarks'.

found on the agricultural land in between. Variability of numbers from both year to year and month to month at both these sites is to a large extent dependent on the differing levels of floodwater, which is the consequence of the vagaries of English weather. The other major development is the excavation of gravel pits across the county (see Fig. 2) which has provided an excellent habitat for diving duck. Two pits in particular, Fen Drayton and Mepal, provide a haven for birds from the Ouse Washes when they are disturbed or when the depth of flooding is insufficient to support them.

We are led to believe that in historic times geese abounded in the undrained fens, but by the time Jenyns began recording numbers of most species were small, this was confirmed more recently by first Evans and then Lack. By the 1950s, however, quite large numbers of Pinkfeet and to a lesser extent Whitefronts were recorded in the central and northern parts of the fens. This proved rather a transient increase, however, and their numbers declined with the agricultural development of the Nene Washes, their major roosting site.

Recently the introduction of feral birds has changed the status of some of the more unusual geese and Canada, Greylag and latterly Egyptian and Barnacle Geese have all begun to appear regularly; the first two species breeding at several sites. Finally in this range of species problems arise with escaped and feral birds. Where the birds are not at present on the British List (e.g. Chloe Wigeon-seen on the Ouse Washes more than once) there is no problem, but in the case of species such as Ferruginous Duck and Red-crested Pochard great difficulty arises in assessing the records. Generally the tameness of the individual, the season of the year, the duration of stay and national occurrences are all taken into consideration in trying to establish the origin of such birds.

Undoubtedly, as there is an increasing level of management on the Nene Washes, and an increase in feral stocks in neighbouring counties, birds in this category are likely to be recorded more regularly, in greater numbers, and in the case of some species establish themselves as breeding species.

SALMON D.G. and MOSER M.E. *Wildfowl and Wader counts 1983-84*. The Wildfowl Trust. 1984

Mute Swan   *Cygnus olor*

Pre 1934

Evans described this species as 'kept on the Cam and elsewhere', while Lack stated that it was common.

1934-1969

Herds gathered regularly on the Ouse and Nene Washes and along the River Cam. The Earith end of the Ouse Washes has always proved very popular, and 400 were recorded there in the winter of 1942 when most of the open water in Cambridgeshire was frozen over. A breeding census in 1955 found 59 pairs in the county. In this period there was a considerable ringing programme, in the Cambridge area alone 201 were ringed in 1960 and 215 in 1961. The recoveries are summarised in Figure 9.

Present Status

A common and widespread resident.

Most areas of water hold at least a pair, and in due season a family. Larger gatherings occur on the washes: Ouse (max. 560 in 1983), Nene (max. 130 in 1983), and Cam (max. 28 in 1960). Non- breeding flocks occur in summer particularly on the Ouse Washes (around 300 in 1983 being the maximum to date). The River Cam in the city of Cambridge has been counted regularly and the winter maxima show a marked decline between 1964 and 1972, possibly due to pollution. In general numbers in winter seem to be increasing on the larger areas of water.

Breeding Status

The first recorded census in 1955 revealed 59 pairs. There was no other

Figure 9. Mute Swan – British ringing recoveries of birds ringed in Cambridgeshire.

count until the national survey in 1978 at which time 44 pairs were counted. Unfortunately, this census coincided with some of the worst floods recorded in May and a number of nests were washed away before counts were finalised. Ogilvie (1981) found that while in general numbers were falling, they were higher in East Anglia although this was largely due to increased numbers of non-breeding birds rather than nests. The census undertaken in 1983 (Ogilvie 1986) confirmed that breeding numbers have changed very little since 1955. While it is possible that there is a decline in the breeding population in the country as a whole it seems stable in Cambridgeshire.

Ringing Results

In the 1960s quite large numbers of Mute Swans were ringed, over 400 in Cambridge city. Brooke (1970), analysing the data showed that a high proportion (over 50%) of deaths, were due to collision with overhead wires, where the cause was known. The recoveries show no particular pattern, but those that were non-local are displayed in Figure 9.

BROOKE M.de L. Some aspects of Mute Swan movement and mortality. *Camb. Bird Club Report* 44. 1970.
OGILVIE M.A. The Mute Swan in Britain 1978. *Bird Study* 28. 1981.
OGILVIE M.A. The Mute Swan *Cygnus olor* in Britain 1983. *Bird Study* 33. 1986.

Bewick's Swan   *Cygnus bewickii*

Pre 1934

Evans described this species as a rare straggler and Lack quoted six records none of which was recent.

1934-1969

Ogilvie (1969) stated that the first influx into Britain occurred in the winter of 1938/39 when birds were driven from the Netherlands by extremely cold weather; thus the regular records of this species began in February/March 1939 with five records of up to 11 on the Ouse Washes. For the following 13 years there were one or two records of up to 26 birds including a record of *C.b.jankowskii*, the eastern race, identified by J.G. Harrison from a shot specimen. In 1953 there was a sudden increase in numbers and a maximum of 102 was recorded on the Ouse Washes with birds on the Nene (5) and Cam (10) Washes as well. There were considerable yearly fluctuations in numbers wintering along the Ouse Washes (see Fig.10) but overall there was a steady increase during this period.

Present Status

An annual winter visitor to the Ouse and Nene Washes and the fenland area in increasing and internationally significant numbers.

The development of the previous period has continued. In 1971 the figure of 1270 on the Ouse Washes in February was a national record since when the figure has advanced to a peak of 6164 in January 1987. Problems have now arisen with accurate counting across the county due to the movement of birds from Ouse and Nene to open farmland feeding sites together with some movements between the two washlands themselves. This makes the total hard to even estimate. Elsewhere, as with the Whooper, this species is rare away from the fenland area but is regularly seen at the closer sites, Wicken Fen, Fen Drayton GP and the Cam Washes, and can be found commonly on the land around and between the two river washes. An attempt has been made to provide extreme dates but injured birds of this species have often spent longer than normal on the washes.

Ringing Results

A bird near the Nene Washes on 28 February 1986 whose ring was read was originally caught and ringed at Vopnafjordur, ICELAND, on 30 July 1984 and had subsequently been seen at Lake Hordasjon, SWEDEN, on 6 April 1986. It was the only Bewick's Swan ringed in Iceland at that time.
Earliest date: 12 October (1974 Ely beet factory)

Figure 10. Bewick's Swan – annual winter maxima on the Ouse Washes 1957-86.

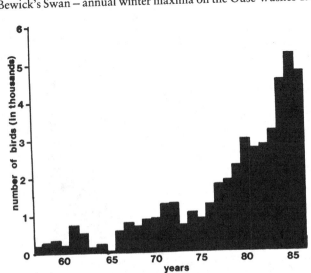

Latest date: 5 May (1974 Ouse Washes)

OGILVIE M.A. Bewick's Swans in Britain and Ireland. *Brit. Birds* 62. 1969.

Whooper Swan   *Cygnus cygnus*

Pre 1934

Jenyns noted that this species occurred frequently in hard winters, while Evans described it as rare, although he added that it sometimes occurred in flocks. Lack quoted very few records of the immediate period of his book and it seems genuinely to have been quite rare.

1934-1969

During this middle period there were annual reports of this species, usually on the Ouse or Nene Washes. The numbers involved were always relatively small (1-30) and there were no extraordinary records, the highest being in 1956 when 27 were seen on the Ouse Washes and 26 on the Nene, but these may have been the same individuals.

Present Status

An annual winter visitor to the Ouse Washes and its immediate area in increasing numbers.

As with Bewick's Swan there has been a dramatic increase in numbers in recent years and the Ouse Washes maxima are presented in Figure 11. It must be noted that, as for many wildfowl, the largest numbers occur in the

Figure 11. Whooper Swan – annual winter maxima on the Ouse Washes 1960-86.

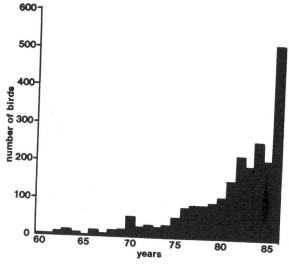

Norfolk section. There have been three significant increases; first in 1970 when the previous 5-20 level jumped to 50 then, after steady stepwise increases, 1981 when there was a jump from 100 to 150 and again the following year to 220 . This level was maintained in the winters 1983/84 and 1984/85 but in 1985/86 a maximum of 330 was counted and in December 1986 520 birds were present, a new record. It remains to be seen where these levels will stabilise but suffice to say the Ouse Washes now contain the only significant numbers of this species in lowland Britain. Increasing numbers on the washes leads to an increasing number of sightings in the area, and places such as Wicken Fen, Fen Drayton GP and the Cam Washes are now regularly visited. Away from this part of the county this species is very rare. On the Nene Washes numbers are beginning to follow the Ouse Washes trend, although they are lower at present, and there will be some interchange. Generally birds arrive in November and numbers are highest in the months from December to mid March before departure at the end of March or beginning of April.

Earliest date: 13 October (1981 Ouse Washes)
Latest date: 13 April (1969 Ouse Washes)

Bean Goose   *Anser fabalis*

Pre 1934

Evans and Lack both seemed certain that this species had occurred but they were uncertain as to detail due to there being confusion in identification between this species and the Pink-footed Goose.

1934-1969

Ennion (*Adventurers' Fen* 1942) claimed a record on Burwell Fen and in 1950 it was said to have occurred in Cambridgeshire but again there were no quoted records. In 1951 there was a record on 14 January on the Nene Washes. In 1961 up to five which were present on the Ouse Washes on 5-6 March were considered to be of the Russian race (*A.f. rossicus*) which local wildfowlers claimed to have seen for several winters! One was seen at the same site in 1969, the first of what has now become an annual visitation.

Present Status

A regular winter visitor in small parties.
From 1970 on the Ouse Washes, where birds from the Norfolk section often wander into Cambridgeshire, it has been noted every winter with the exception of 1977 and 1983. There are usually 1-2 birds but in 1979 a

party of 40 flew over the Ouse Washes. Away from the Ouse Washes and its environs it is extremely rare.

Earliest date: 20 November (1979 Ouse Washes)

Latest date: 24 March (1978 Ouse Washes)

Pink-Footed Goose   *Anser brachyrhynchus*

Pre 1934

Evans stated that it 'occurs at times' but he suggested some confusion with the Bean Goose. Lack described it as a regular winter visitor to the county. He described 'large flocks' in the extreme north of the county, and stated that small parties were 'not infrequent' on the Ouse Washes. A single event in November 1918 is noted when Herdman (*Brit. Birds* XII 1919) described a 'huge passage of Grey geese, presumed to be Pinkfeet, over Cambridge on a misty night.'

1934-1969

In the first part of this period, up to the early 1950s, there were records of small parties mainly on or around the Ouse Washes area. From 1953 details of larger numbers were published (up to 500 on the Ouse Washes) and this situation continued with flocks of several thousand roaming the northern and central fens until in 1963 agricultural development of the Nene Washes seems to have instantly reduced their status to its former level.

Present Status

A fairly regular winter visitor in small parties.

Numbers up to 150 have been recorded, but normally from 1 to 30 are seen. The greater numbers are found around the Nene Washes and the smaller in the general area of the Ouse Washes. Away from these two areas this species is rare and the only recent record for the southern half of the county is of a bird flying over Fowlmere in January 1976. To date the number of escaped birds appears to be very small and there is no known local feral population.

Earliest date: October (1945 Stretham)

Latest date: March (1967 Ouse Washes)

White-Fronted Goose   *Anser albifrons*

Pre 1934

Jenyns noted that this species was not very common while Evans went further and described it as rare. Lack quoted only five records but stated that Farren reported to him that among wildowl shot on the Washes, and for sale on Cambridge market, were one or two Whitefronts.

1934-1969

There was a flock of 50-60 on Cottenham Fen around Christmas 1938, one of which was later shot; however, this was a rather high count since numbers around this time were usually of the order 1-20. Generally it seems that there were records in most years on or around either the Ouse or Nene Washes. In the later 1950s and early 1960s larger parties were reported at times; e.g. 90 on the Nene Washes in 1959 and 60 on the Ouse, or in 1960 200 on the Nene and 80 on the Ouse. Following these years numbers were lower.

Present Status

An uncommon but annual winter visitor on, or around, the Ouse Washes. As in the previous period, with a few exceptions, there have been parties of 1-20 on the Ouse or Nene Washes almost every winter with birds flighting and feeding in the immediate vicinity. In January 1970 a maximum of 140 were counted on the Ouse Washes. Elsewhere it is almost unrecorded, and those few records probably refer to escaped or feral birds.

Earliest date: 30 October (1976 Ouse Washes)
Latest date: 16 April (1961 Ouse Washes)

[Lesser White-Fronted Goose   *Anser erythropus*

One at Pymore on the Ouse Washes on 14 October 1968 was thought to be an escape.]

Greylag Goose   *Anser anser*

Pre 1934

Evans stated that bones of this species have been found in the peat in the fen districts and that there is evidence that it bred until 1773. Lack repeated Evans' information adding that there were no confirmed records since 1800.

1934-1969

The first modern record is of a single bird on or around the Ouse Washes about Christmas 1944. In 1950 a record of six shot from a flock of 300 caused the *Camb. Bird Club Report* editor to express surprise at the paucity of records since wildfowlers encountered the species regularly along the Nene Washes. Later four seen there on 21 February 1954 were said to be the first convincing record for the county! There followed a spattering of records, usually of single birds on or around the Ouse Washes, until in 1961, when there was the first mention of a feral bird. From 1959 on there have been records each year and they have involved more and more birds staying for longer periods mainly as a result of introductions and wandering feral groups.

Present Status

An increasing resident, due to expansion of the feral population with some wild birds visiting in winter.

A relentless advance of the feral stock has taken place over the last decade, based mainly on the Ouse Washes, and consequently birds of this species can be seen around the county at almost any gravel pit or wetland. There are large gatherings of up to 300 birds on the Ouse Washes in winter and significant numbers at Fen Drayton GP and Wicken Fen. Elsewhere numbers are usually smaller with family parties most common.

Breeding Status

The feral stock appear to have first bred at Wicken Fen in 1981; they are, however, already expanding with nesting now recorded at both Wimblington and Waterbeach GPs as well.

Canada Goose   *Branta canadensis*

Pre 1934

While unmentioned by both Jenyns and Evans, Lack stated that this species was 'reported as an escape not infrequently from the Ouse Washes.' The only records traced are three before 1843 and a flock of 20 on the Ouse Washes in December 1929.

1934-1969

Up to 1957 there were only two more records, in 1954 and 1955. In 1957 the introduction of birds at one or two sites, and incursions into the county, produced the first regular records, and in 1959 the first breeding was recorded at Milton GP. This was followed by other pairs setting up at Shepreth and Hauxton GPs in 1961. By 1965 there were large gatherings at several sites: 60 at Harlton, 34 at Great Shelford (both presumably from the Hauxton-based flock) and 40 at Shepreth, and in the winter of

1968/69 up to 150 were roosting at Hauxton. By 1970 Hauxton had a breeding population of seven pairs and birds were being seen across the whole of the southern part of the county as far north as Wicken Fen. The speed of their expansion had been very rapid. Limentani (1974) showed from ringing at Hauxton that in the early 1960s the birds that bred there wintered in north Norfolk at Holkham; whereas after a period of wintering on their breeding sites in the late 1960s they began to move south-east on a line towards the Blackwater Estuary.

Present Status

A moderately common and widespread breeding resident found on most suitable areas of open water.

From the early 1970s to the present day numbers have continued to increase. Breeding success appears to have been good and this has resulted in many roving parties, particularly in autumn. The area from which records are received is still however predominantly the Ouse Washes southwards. A clear pattern of winter movements seems to have ceased with birds flighting from site to site according to weather conditions and disturbance.

Breeding Status

After the first success in 1961 numbers rose rapidly and by 1967 23 pairs were breeding at five sites: Hauxton GP, Hatley Hall, Waterbeach GP, Wicken Fen and Chittering. By and large these sites, with the exception of the last, held the nucleus of the population until 1977 when the Hatley flock was deliberately eliminated! In recent years numbers at the former breeding sites have remained fairly stable but there has been a further expansion into the gravel pits that were dug during the building of the M11 (e.g. at Impington and Whittlesford).

Ringing Results

Two birds ringed at Hauxton in June 1972 and 1975 were recovered at Stansted (culled) and Regent's Park, London, in December and October 1976 respectively.

LIMENTANI J. Changes in the wintering habits of Canada Geese. *Camb. Bird Club Report* 48. 1974.

Barnacle Goose   *Branta leucopsis*

Pre 1934

Evans stated that it was 'said to have been obtained in the fens and near the coast'. Lack quoted two records both from the nineteenth century.

1934-1969

There were five more records in this period all of which probably refer to wild birds mainly on the Nene Washes where a flock of 15 were recorded in January 1951. This made a grand total of nine records by 1970 indicating considerable rarity.

Present Status

An extremely uncommon winter visitor whose status is greatly complicated by the emergence of an escaped and feral population. As with other geese the status of this species has changed dramatically since 1970 with the introduction of feral and escaped birds. Since 1973 there have been records almost every year and these are by no means confined to winter months although it is possible that the same bird may be involved on some occasions. To date records still refer only to single birds many of them are associated with other geese on gravel pits and similar sites. One, introduced as an injured bird in 1982, lives along the River Cam between Cambridge and Fen Ditton. Others, more certainly of feral origin, have taken up residence at Waterbeach and Shepreth GPs and it is likely that this species will increase and perhaps eventually breed in the county.

Brent Goose   *Branta bernicula*

Pre 1934

Jenyns noted that birds of this species had been shot in Cambridgeshire, and Evans thought it under-recorded. Lack quoted six records between 1912 and 1933 and also mentioned that Farren had seen them for sale on Cambridge market (see White-Fronted Goose).

1934-1969

There were nine records in this period, mostly from the Ouse or Nene Washes.

Present Status

An uncommon but probably increasing winter visitor.
A further 13-14 records in 16 years probably reflects the rise in numbers around the east coast as well as increased observation, and with eight records in the years 1980-85 it is possible that regular visits will become more commonplace as with other members of this genus.
Numbers: usually only single birds are seen but on one or two occasions up to seven have been reported. The favoured site remains the Ouse Washes with others recorded at Wicken Fen, Fen Drayton and Waterbeach GPs and on the Nene Washes. The monthly distribution of all the records appears in Figure 12.
Earliest date: October (1912 Royston)
Latest date: 2 May (1982 Ouse Washes)

[Red-Breasted Goose   *Branta ruficollis*

Not mentioned by Evans but Lack, quoting Yarrell (1837) records one or more killed by Stephens in the severe winter of 1813. Lack goes on to point out that Howard Saunders (1889) makes no mention of this record although he states that unsubstantiated claims were made, in which category Saunders may well have included the above.]

Egyptian Goose   *Alopochen aegyptiacus*

Pre 1934

Lack stated that this species was reported as an escape.

1934-1969

The first documented records follow annually from 1964. During this initial period there were five records mostly on or near the Ouse Washes

Figure 12. Brent Goose – monthly distribution of all records.

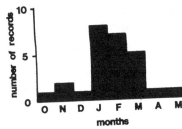

with a particular preference for the Earith area. Most were of single birds though on one occasion up to three were seen. There was no particular seasonal pattern.

Present Status.

An unusual, but increasing, visitor, seen in all months of the year. Although these birds breed in neighbouring Norfolk there is as yet no sign of significant increase of occurrence in Cambridgeshire. There have been 16 records between 1970 and 1984, almost annual, but usually of single birds on a single occasion. Exceptions were in 1974 when 1-2 birds stayed on the Ouse Washes from 24 February to 6 June, two males and a female at Wicken Fen in January 1980 and a bird that made numerous appearances at Whittlesford GP throughout 1987. The inescapable conclusion is that most of these birds are simply visitors from the Norfolk stronghold.

Ruddy Shelduck   *Tadorna ferruginea*

Mainly occurs as an introduced, feral species, being native in Africa and Asia. There is also a population in southern Europe.

Pre 1934

Lack reports that Farren received several specimens, including one from Cambridge sewage farm, all presumed to have escaped from collections.

1934-Present Status

All the available records are listed:

One on the Ouse Washes, 14 October 1968.

One on the Ouse Washes, 16 January 1978.

One at Cambridge sewage farm, said to be obviously feral, 9 September 1979.

One at Ely beet factory, 11 August 1984.

One at Fen Drayton GP, 24 September 1984. (This, and the previous bird may have been of Dutch origin since there were several on the coast at this time.)

A single on the River Cam near Chesterton, 3 April 1985.

A female at Wicken Fen on 5-9 April and 24 May 1985 arrived following a spell of strong south-westerlies (that brought a number of Saharan migrants).

A female flew off to the west from Fen Drayton GP, 25 May 1985.

Shelduck *Tadorna tadorna*

Pre 1934

Evans described this species as an occasional winter visitor. Lack reported irregular occurrences on the Ouse Washes, at Fulbourn Fen and Cambridge sewage farm, 1-2 nearly every year either on spring or autumn passage, or, erratically, during winter.

1934-1969

A considerable change in status took place during this period. The number of sightings increased, and breeding was recorded on the Nene Washes (c.1936), at Peterborough sewage farm (from c.1945) on the floods at Haddenham in 1947 and possibly at Cambridge sewage farm (1951 and 1952). A pattern of regular breeding emerged on the Nene Washes with up to 14 pairs, while on the Ouse Washes nesting began about 1969.

Present Status

Recorded throughout the year. A winter visitor, passage migrant and breeding species.

In the past 15 years numbers have continued to increase. On both the Ouse and Nene Washes numbers of over 100 are regularly recorded, usually between February and May. One or two birds are recorded annually at other local wetland sites, e.g. Ely beet factory and Wicken Fen.

Breeding Status

Nene Washes

While early records suggested that breeding was first recorded in 1947, it was later discovered that this species had bred on the Nene since 1936 (Nisbet and Vine 1955). Numbers built up very gradually but it appears that in the 1970s nesting did not take place in some years. The maximum number of pairs was 20 in 1980.

Ouse Washes

Regular breeding began in 1969 and has taken place almost annually since with the number of pairs varying from 2 to 51.

Ely beet factory

Breeding began in the early 1970s but has continued in a rather spasmodic fashion. The maximum number of pairs recorded was two in 1985.

Fen Drayton GP

Two pairs bred in 1984 and were present in 1985.

Cam Washes
A pair bred in 1984 and were present in 1985.
Wicken Fen
A pair was present in 1985.

NISBET I.C.T. and VINE A.E. Regular inland breeding of Shelducks in the fens. *Brit. Birds* 48. 1955.

Mandarin Duck   *Aix galericulata*

A feral species of Asian origin.
Several have occurred in Cambridgeshire. Apart from a few Ouse Washes records, e.g. 26 April 1975 and 26 August 1979, these seem to have been escapes from nearby collections. The male on the River Cam in the 1970s for instance was doubtless from the Botanic Garden.

Wigeon   *Anas penelope*

Pre 1934

Lack described this species as the commonest duck on the Washes. The numbers that he quoted are small in comparison with the present time, the

Figure 13. Wigeon – annual winter maxima on the Ouse Washes 1960-85.

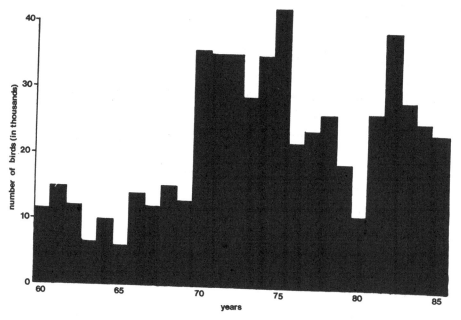

maximum reported being around 1000 which presumably refers to birds counted at either Earith or Welney since no complete counts were made at this time.

1934-1969

Disturbance from shooting and the bombing range on the Ouse Washes kept numbers low during the first part of this period but it seems that there was considerable interchange with south Wash populations with daily flights between these localitiies. On the earliest complete count in 1951 there was a maximum of 5000, which far exceeded previous records, and in March 1955 the numbers reached a new peak of 19 500, after which they remained in the region of 10 000 birds each winter except in years when the floods were reduced. On the Nene Washes counts varied from 200-300 to 2500 in 1961, and a maximum of 12 000 in early March 1969. At Wicken Fen the excavation of the new mere produced annual maxima of 400-500 and exceptionally 1000 in December 1960. During this period records of summering birds on the Ouse Washes began although it does not seem as though any of these birds attempted to nest.

Figure 14. Wigeon – build-up of numbers on the Ouse Washes over the winter 1982/83.

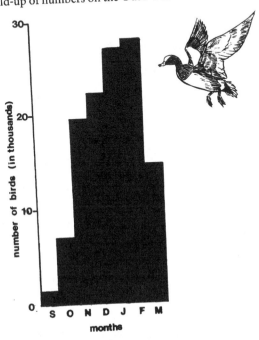

Present Status

A winter visitor, passage migrant and, in small numbers, summering bird with breeding suspected but never proven.

In the last 15 years on the Ouse Washes new peaks, more than double those of the previous period, have been recorded with an all-time maximum of 42 500 counted in February 1975 (see Fig. 13). (As with many species the majority of these birds were to be found on the Norfolk section in the Wildfowl Trust reserve.) The other sites where large numbers winter are the Nene Washes and Wicken Fen. On the Nene the numbers have generally increased with a maximum of 6680 in 1977, and at Wicken Fen/Cam Washes peaks of 1000 birds are counted in most winters. Almost all of the larger gravel pits are visited at times, with Fen Drayton, Landbeach and Waterbeach being most favoured. There is a marked seasonal build-up of numbers with the peaks being reached in February or March and Figure 14 shows the monthly counts over the winter 1982/83. This build-up may be associated with both the extent of the flooding on the Washes, and the arrival of passage birds. Most birds leave by the end of April and return, with a gradual build-up from October through November, to large numbers by mid December. Oversummering has become more common on the Ouse Washes with breeding often suspected but as yet not proven.

American Wigeon   *Anas americana*

An accidental of North American origin. Two records, listed below:

1 A male on the Ouse Washes on 12 February 1956 could have been an escape.
2 A male, on the Ouse Washes, north of Mepal, 20-22 March 1980.

Gadwall   *Anas strepera*

Pre 1934

Jenyns gave several winter records from Burwell, Cottenham and Willingham. Evans described it as 'occasionally met with'. Lack stated that by 1934 its status was that of an irregular visitor. There was a single breeding record from 'near Wicken Fen' the nest being 'taken' in 1917 and the eggs examined by Jourdain who considered them to be of this species.

1934-1969

During the middle period records were rather sparse and it seems possible that it was under-recorded; generally there were only 3-5 records per annum. Breeding was proven at Burwell Fen in 1938 and on the Ouse Washes in 1953 and 1964. Easy (1982) stated that birds summered on the Nene Washes in the early 1950s and these, plus birds at Peterborough sewage farm, could have resulted from introductions, or immigration from the Continent. It is possible that some of these birds bred.

Present Status

Mainly a winter visitor with some passage birds. Breeds regularly at some sites.

In recent times this species has expanded and birds are now seen at many of the open waters in the county, e.g. Cam, Ouse and Nene Washes, Ely beet factory, Eldernell, Fen Drayton, Landbeach, Milton and Waterbeach GPs. Easy (1982) considered it likely that these birds originated from the Breckland stronghold. Winter numbers on the Ouse Washes reached a peak of 390 in 1979, but in most winters around 100-200 are present. Elsewhere numbers are generally small.

Breeding Status

In the 1960s breeding on the Ouse Washes became a regular occurrence reaching a peak of 20 pairs by 1969 in which year a pair also nested on the Cam Washes. Numbers increased in the 1970s to a maximum of 52 pairs in 1974 and remained around 30-50 until 1980 when only three broods were located. In some years (e.g. 1985) large numbers of males are present (76) but the number of broods raised is nevertheless small. Flooding also affects the nesting success on the Ouse Washes. In the 1970s breeding was also discovered at Ely beet factory, Fulbourn Fen, Kennett GP, Nene Washes, Waterbeach GP and Wicken Fen, with usually only one or two pairs at each site.

Ringing Results

A bird ringed at Aznalcazar, Sevilla, SPAIN, on 25 June 1968 was found on the Ouse Washes on 24 January 1970.

EASY G.M.S. The Gadwall - A success story. *Nature in Cambridgeshire*. 25. 1982.

Teal   *Anas crecca*

Pre 1934

Evans described this species as 'somewhat rare, probably breeding'. Lack stated that it bred at both Chippenham and Wicken Fens as well as at other sites. In winter it was a common bird on fens and washlands occurring in flocks of several hundred. It was also found in smaller numbers elsewhere in the county.

1934-1969

Up to 1954 the status of this species differed little from Lack's description. In February of that year over 2000 were present on the Ouse Washes and in 1955 an all-time high of 4200 was counted. This was the beginning of much larger numbers wintering there regularly. On the Nene Washes a report of 2500 in early March 1969 far exceeded any previous count for the site.

Present Status

Mainly a winter visitor, with some passage, and a small number of summering birds. Breeds in small numbers at selected sites.

Winter maxima on the Ouse Washes have usually been between 1000 and 4000 and exceptionally 7570 in 1975. On the Nene Washes counts have peaked at 2400 (1980), although these high numbers are relatively recent. At Wicken Fen and Ely beet factory numbers away from the breeding season vary between 100 and 900 with usually around 300 at the former and 100-200 at the latter. Elsewhere this species is found in small numbers at most suitable sites and can occur anywhere in the county on any small patch of water. At Fulbourn Fen it seems to have declined but it is still present in some numbers on Chesterton Fen in most years. There have been records, particularly in recent years, of the North American race (Green-winged Teal, *A.c.carolinensis*) on the Ouse Washes.

Breeding Status

Recorded regularly on the Ouse Washes where the number of pairs is rarely more than 15. At Wicken Fen pairs are thought to attempt breeding in most years and were regularly recorded in the 1960s and 1970s (Thorne and Bennett 1982). Elsewhere the situation is not well documented although there are suitable sites (Cam Washes, Fulbourn Fen, Nene Washes, etc.) where birds are reported in summer without there being real evidence of successful nesting.

Mallard   *Anas platyrhynchos*

Pre 1934

Surprisingly, Evans described the Mallard as 'not very common'. Lack, however, stated that it bred fairly commonly and that there were some notable autumn influxes. He also noted that numbers in winter exceeded those in summer and that the species was more generally distributed than other duck. Flocks of several hundred were said to be not uncommon.

1934-1969

In the middle period numbers increased so that by the 1950s over 1000 birds regularly wintered on the Ouse Washes, and by 1960 the peak numbers often exceeded 4000. At most sites autumn flocks of several hundred were a regular occurrence.

Present Status

A common and widespread resident, and winter visitor with some through passage.
Found throughout the county, usually in small numbers, often just a pair. On the Ouse Washes the wintering maxima have varied between 1540 (in

Figure 15. Mallard – annual winter maxima on the Ouse Washes 1959-85.

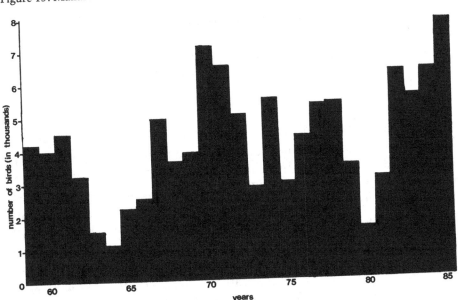

March 1980) and 7815 (in December 1985). On the Nene Washes there are usually up to 1000 present in the winter, and at Wicken Fen the numbers have increased from a regular 300 in the 1970s to nearer 700 in the 1980s. Along the River Cam in Cambridge the maxima vary between 650 and 700 although exceptionally there were 827 counted in January 1980. The larger Washes counts often reflect periods of intense agricultural activity which drives birds from the surrounding fenland arable to the refuge of the washland system.

Breeding Status

This species breeds commonly and in the *Breeding Atlas* (Sharrock) was recorded in most 10 km squares, but often the number of birds at any one site is merely a pair or two. On the Ouse Washes there were up to 1000 pairs in the early 1970s but there are fewer at present.

Ringing Results

A bird ringed in August 1970 at Wicken Fen was shot at March in October of 1973. A bird ringed at Deeping St James (Lincs) in September 1974 was killed by a fox at Wicken Fen in August 1978. Finally, a bird ringed at Wicken Fen in October 1961 was found near Calais, FRANCE, in September 1963.

Pintail *Anas acuta*

Pre 1934

Evans described this species as 'a rare straggler in winter'. Lack stated that numbers of up to 12 were seen annually on the Washes in winter and that a few were seen in the Burwell/Wicken area. He also stated that there was a possible breeding record at Earith in 1927.

1934-1969

By 1947 several hundred were visiting the Ouse Washes each winter and in the 1950/51 winter the first count of over a thousand was recorded on the Nene Washes. At this time there were also regular records of small numbers at both Cambridge and Peterborough sewage farms and Wicken Fen. Nisbet (1954) stated that large numbers rarely arrived before December and that the peak period was during February and March. After 1950 numbers on the Ouse Washes fluctuated considerably for a while but eventually four-figure maxima became standard, with over 2000 counted on 7 March 1954, and 4000 in March 1957. In 1969 5000 were estimated to be on the Nene Washes in early March and a further 1200 were counted along the Ouse Washes, making a total of over 6000, by far the largest number to have been reported in the county.

Present Status

An unusual winter visitor and rare summering/breeding species. Found mainly on the washlands of the Ouse and Nene. After a long period in the 1970s when up to 3000 were present on the Ouse Washes, and several hundred were on the Nene Washes, the situations reversed so that in 1980 there was a maximum of 400 at the former site and a maximum of 2700 at the latter. More recently there has been a tendency for this to even out with maxima of 1000 and 1700 respectively in 1983. As with other duck wintering on the Washes, birds of this species occur in significant numbers (up to 50) at other nearby sites, notably Ely beet factory, Fen Drayton GP and Wicken Fen. Smaller numbers (usually singles or pairs) can be seen at sites such as the Cam Washes, Fulbourn Fen, Milton GP, Landbeach GP, Cherry Hinton Cement Pits and Waterbeach GP.

Breeding Status

The first reported records (1927 and 1928) were based on circumstantial evidence but breeding was finally proved in 1947 on the Ouse Washes and on the Nene Washes in 1951. Since this time breeding has occurred sporadically at both sites.

Ringing Results

A bird ringed in November 1972 at Sonderho Fano, DENMARK, was found at Manea (Ouse Washes) the following January, giving a sole clue as to the origin of these birds.

NISBET I.C.T. The status of Pintail inland in the fens. *Brit. Birds* 47. 1954.

Garganey *Anas querquedula*

Pre 1934

There are only four records from the nineteenth century indicating that it was very rare at that time. Evans described this species as a 'rare summer or autumn visitor'. Lack stated that it was a summer visitor breeding regularly at 1-3 sites and that there was a late summer influx. He concluded that there were signs of increase.

1934-1969

In the early part of this period numbers were still very small, but in 1952 10-14 pairs were said to have bred on the Nene Washes and 25-35 pairs on the Ouse Washes (23 pairs were present there in 1967). These numbers seem to be exceptional although in 1954 an autumn flock of 60 appeared on the Ouse Washes. Breeding was also recorded at Chesterton Fen (1954) and at Fulbourn Fen (1957, 1959). In the 1960s numbers seem to have declined to their former levels.

Present Status

A regular summer resident in very small numbers to traditional sites. Some autumn passage.

Usually recorded from April to September, breeding almost annually on the Ouse Washes with normally 4-6 pairs present. Other sites are in regular use, e.g. Cam Washes, Nene Washes and Wicken Fen, although the number of birds is usually very small. There is a tendency for numbers to build up in autumn, although it is not clear whether this is due to congregations of local birds and their progeny, or to an influx from outside the area. Other sites in the county, where there is suitable water, are often visited.

Earliest date: 3 March (1984 Ouse Washes)
Latest date: 11 October (1975)

A few pairs of Garganey nest most years on the Ouse Washes.

Blue-Winged Teal *Anas discors*

An accidental of North American origin and an escape from wildfowl collections.

Three records, listed below:

1 A female, on the Ouse Washes between Purl's Bridge and Welney, from 11-14 October 1974.
2 A male, on the Ouse Washes near Over, from 18-22 February 1978. (This bird probably came from Fenstanton GP, Hunts where a bird of this species was present on 11-12 February.)
3 A male on the Ouse Washes near Purl's Bridge, 22-24 June 1983.

Shoveler *Anas clypeata*

Pre 1934

Jenyns knew this species only as a winter visitor. It appears to have first bred in the county at the turn of the century, having begun to increase as a breeding species in Britain around 1876. Evans stated that it occurred from time to time, possibly breeding. At the time of Lack's review it was a regular winter visitor on the Ouse Washes in small numbers, breeding at Burwell, Fulbourn and Wicken Fens.

1934-1969

In the middle period there were regular sightings at the usual wetland sites, with breeding recorded in most years. The floodwater of Burwell Fen proved especially attractive and 20 pairs were present there in 1936. By the 1950s there were significant numbers nesting on both the Ouse and Nene Washes (30 pairs at each in 1951), and in 1952 there was an increase to 120 pairs on the Ouse Washes with other breeding records from Clayhithe, Hauxton and Wicken Fen. During this period winter maxima varied between 100 (1954) and 950 (1961) on the Ouse Washes, and by the 1960s there were over 100 recorded at both the Nene Washes and Wicken Fen in most winters.

Present Status

Present throughout the year. Some passage birds, more in spring than autumn, and some wintering influx. Breeds at several sites and is probably increasing.

Found on the Ouse, Nene and Cam Washes and at Wicken Fen and Ely beet factory. Up until 1979 winter numbers were higher on the Ouse

(500-600) than the Nene (up to 100), since when they have evened out at around 300 at each site. In some years there are large gatherings at both Ely beet factory (up to 130) and at Wicken Fen (up to 120), but more usually each of these sites holds about 10-20 birds. Other sites where this species has been recorded are Chesterton Fen, Fen Drayton GP, Land-beach GP, Milton GP and Waterbeach GP.

Breeding Status
On the Ouse Washes up to 300 pairs nest with varying success. On the Nene Washes the numbers are usually much smaller with a maximum of 60 pairs in 1980. Breeding also takes place regularly at Wicken Fen and Ely beet factory and has been recorded at Fen Drayton GP.

Ringing Results
A bird ringed on 9 December 1959 at Lekkerkek, NETHERLANDS, was recovered eight days later at Littleport, and one found near Mepal in January 1979 had been ringed at Slonsk, POLAND, the previous July.

Red-Crested Pochard   *Netta rufina*
Pre 1934
Evans made no mention of this species. Lack stated that Millais (1913) possessed an adult male that had been killed near Cambridge in the winter of 1882 and saw others killed at Ely on the same day.

1934-1969
Only four records:

An adult male at Chittering on 14 February 1954. (This bird coincided with an influx of diving duck from the Continent during a cold spell and was therefore believed to be wild.)

Two immature females were at Milton GP from 6 October to 5 November 1964.

Two records of two birds, of feral origin on the Ouse Washes on 30 January and 15 May 1966.

Present Status
A vagrant from the Continent whose true status is greatly complicated by the occurrence of both escaped and feral birds.

There has been a marked increase in records with at least ten in the 1970s and 14 between 1980 and 1987. At Fen Drayton some of these birds stayed for long periods particularly in the 1980s. In recent years there has also been more than one bird at some sites, e.g. two males and a female at Fen Drayton GP from July to September 1985 and again in 1987 when two males were present in July. It is thought that the October and November sightings are of genuinely wild passage birds as are those in January and February which are almost certainly birds displaced from the Continent during spells of harsh weather. For the rest it is more than likely that their origins are feral or wildfowl collections.

Pochard *Aythya ferina*

Pre 1934

Evans, with some foresight, described this species as a winter visitor and a possible future breeding species. Lack reported that it was a regular visitor in small numbers on the Washes with seldom more than 20 in a flock. Larger numbers, he suggested, were associated with hard weather.

1934-1969

In 1934 Evans prediction was fulfilled and breeding was recorded at two sites; one pair at Wimpole and another at Burwell (Vine 1961). In the 1940s wintering numbers began to increase so that by 1947 there were up to 200 on the Ouse Washes. This increase continued apace and within

Figure 16. Pochard – annual winter maxima on the Ouse Washes 1967-85.

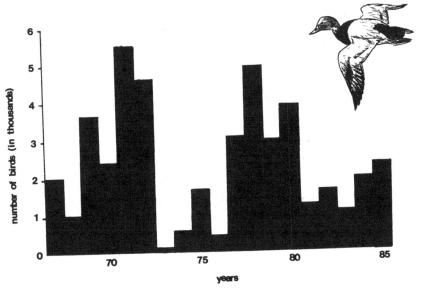

four years (1951) there was a record of 1000 on the Ouse Washes in February together with 125 at Ely beet factory. Thereafter throughout the period, along the Ouse Washes, most winter maxima were around the 700-800 mark, with the exception of February 1969 when a further record number of 3650 were counted.

Present Status

Mainly a winter visitor with a small breeding population.

By far the largest winter counts are on the Ouse Washes (see Fig. 16) where up to 5000 gather in some winters. Elsewhere numbers are fairly consistent: Fen Drayton GP (up to 300), Landbeach GP (up to 100), Milton GP (up to 150), Mepal GP (up to 900), Nene Washes (up to 250), Cherry Hinton Cement Pits (up to 200) and Waterbeach GP (up to 250). Smaller numbers (less than 200) can be seen on the other areas of open water.

Breeding Status

Nesting was first recorded in 1934, but was subsequently rather spasmodic until the early 1950s when regular breeding began at Hauxton. Records are now received from both the Ouse (1-3 pairs) and Nene (1-2 pairs) Washes; elsewhere breeding attempts are suspected with pairs present throughout the summer at Hauxton, Landbeach, Milton, Waterbeach and Wimblington GPs, Wicken Fen and Ely beet factory, but success has been very limited.

VINE A.E. The breeding status of three diving species in the last fifty years. *Camb. Bird Club Report* 35. 1961.

Ring-Necked Duck   *Aythya collaris*

An extremely rare accidental from North America.

Three records, listed below:

1 A single bird at Mepal GP, 7-10 January 1968.
2 A male at Milton GP, 28 March-22 April 1976 and almost certainly the same bird at Landbeach GP, 24-30 April.
3 A male, that had been at Earith GP (Hunts) from 23 March to 29 April 1977 was seen once at the Earith end of the Ouse Washes.

Ferrugnious Duck   *Aythya nyroca*

As a whole the status of this species is extremely difficult to report due to the unsatisfactory nature of the nineteenth-century records. Therefore all sightings quoted in the *Camb Bird Club Reports* are listed but without being numbered.

Pre 1934

One, around 1830, in Thackeray's collection came from Cambridge Market [Lack]. (This implies it originated in the county but that is by no means certain).
[Lack noted that Howard Saunders referred to more than one specimen.]
One at Hauxton GP, 10-12 October 1954.
One at Milton GP and latterly at Wicken Fen, 16 January-2 April 1960.
A drake at Waterbeach GP, 21 February 1970.
A drake on the Ouse Washes, 19 October-21 November, and at Mepal GP, 24 November 1974. This bird might have been an escape.
A drake at Milton GP, 6 April 1975.
One, thought to be a juvenile, at Waterbeach GP, 3 August 1976.
One adult male at Waterbeach GP, 14 September 1976.
A drake at Cherry Hinton Cement Pits, 17-24 January, and (another) on 29 April 1984.
One at Ely beet factory, 17 September and 14 October 1984.

Tufted Duck   *Aythya fuligula*

Pre 1934

Evans described it as being similar in status to the Pochard, thereby suggesting its development as a breeding species. Lack was able to confirm the prophecy with breeding records at Arrington in 1911 and possibly subsequently; he also noted a similarity in status with the Pochard.

1934-1969

Further breeding records at Burwell (1937) and Wimpole (1941) indicated an expanding population, and by 1947 records of 200-300 were being reported from the Ouse Washes during the winter months. By the mid 1950s there were more than 10 pairs at 3-4 sites annually (Vine 1961) and by 1967 this had risen to 46 pairs at 11 sites. Wintering numbers on the Ouse Washes remained rather variable but maxima of 600 in 1966 and 620 in 1967 were record totals.

Present Status

Mainly a winter visitor, with a widespread, stable, but modest breeding population.

Found on most gravel pits and areas of open water in winter. Highest aggregations are: Ouse Washes (up to 1100), Wicken Fen (up to 300), Whittlesey-Nene Washes (up to 230), Fen Drayton, Landbeach, Milton, Roswell and Waterbeach GPs (all up to 200). A marked increase in numbers was noted in 1973 and this has been maintained. Results from the wildfowl counts indicate that the annual wintering population averages around 1000 birds with about half of these on the Ouse Washes/Mepal GP area and the rest evenly distributed around the various gravel pits, i.e. Barrington, Hauxton, Wimblington and Whittlesford in addition to those mentioned above.

Breeding Status

The number of breeding pairs has not changed dramatically over the last 15 years and is within the range of 35-60 successful pairs; again about half of these are on the Ouse Washes with the remainder at the gravel pits. In 1977, when the largest total was recorded on the Ouse Washes in winter (1100), around 150 pairs attempted to breed.

Ringing Result.

A bird ringed at Ely beet factory in August 1984 was found at Ijselmeer, NETHERLANDS, the following December.

VINE A.E. The breeding status of three diving species in the last fifty years. *Camb. Bird Club Report* 35. 1961.

Scaup *Aythya marila*

Pre 1934

Jenyns listed three records all between 1827 and 1830. Evans described this species as 'reported from the fens and elsewhere in winter', which rather lacks precision; however, by the number of records quoted by Lack it was clearly a rare visitor.

1934-1969

Up to the end of the war there were perhaps no more than five or six records, the most exceptional being of up to 10 birds near the Earith Washes. Post-war records, however, show a more regular pattern and from 1953 the species has been recorded regularly with between one and seven records per annum. The more spectacular sightings include 60 on the Ouse Washes in 1947, and 20, 19 and 16 in the years 1963, 1966 and 1967. The only summer record was on 3 July 1949 when a female was seen at Cambridge sewage farm but it seems highly probable that this bird was an 'escape'.

Present Status

An unusual but regular winter visitor.
In most years there are one or two records, usually in the period November to March, involving single birds or pairs. The site visited most regularly is the Ouse Washes, although occasionally birds are recorded at sites such as Ely beet factory, Fen Drayton, Landbeach, Milton or Waterbeach GPs, the River Nene and Wicken Fen. A single male was present at Milton GP from late 1980 to May 1982, with a short break in January 1982. This bird may well also have been an escape (see above).
Earliest date: 15 September (1971 Milton GP)
Latest date (excluding summer bird): 30 April (1976 Milton GP)

Eider *Somateria mollissima*

An extremely rare storm-blown vagrant.
Four records, listed below:

1 An immature, in the University Museum, Department of Zoology, labelled 'Cambridgeshire'. [Lack]
2 130 flew over Pymore (Ouse Washes) on 11 November 1975. (This record, which is unprecedented, occurred at a time when there was severe fog on the coast.)
3 One shot on the Nene Washes in December 1977.
4 A male, in summer plumage, on Cherry Hinton Cement Pit on 18 February 1981.

Long-Tailed Duck *Clangula hyemalis*

A very rare vagrant.
Eleven records, listed below:

1 Jenyns recorded one shot near Ely in 1827 and found in the collection of Dr Thackeray. [Lack]
2 Jenyns also noted one from Ely in the Cambridge University Museum of Zoology. No date. [Lack]

3 One near Wisbech. No date. [Evans]

4 Two immatures at Cambridge sewage farm, 8-16 November, 1946 after a strong north-easterly wind.

5 An adult male at Cambridge sewage farm, 29 June 1947, after a thunderstorm. (It is possible that this bird was an escape, the month of its occurrence makes it an extremely suspicious record.)

6 A male on the River Ouse near Over, 26 April 1953.

7 A female at Cambridge sewage farm on 26-27 October 1957 was last seen on the River Cam nearby.

8 An immature at Milton GP on 9 November 1967.

9 One at Ely beet factory from 23-27 October 1974.

10 A female/immature at Fen Drayton GP, 24 January 1981.

11 A female at Fen Drayton GP, 23 April 1983.

Common Scoter   *Melanitta nigra*

Pre 1934

Jenyns, on the authority of Dr Thackeray, noted large flocks in the fens, but Lack commented that he considered this unlikely. Lack also listed six records, the last of which is dated 1931.

1934-1969

During this period records, while remaining irregular, increased in number considerably. There were six records in the 1940s, fourteen/fifteen in the 1950s, mostly at Cambridge sewage farm/Milton GP, and ten in the 1960s, of which six were at Milton GP.

Figure 17. Common Scoter – monthly distribution of all records.

Present Status

An uncommon winter and passage visitor.

In recent years the previous trend has continued and there were a further seven records in the 1970s and four in the 1980s up to 1985. More sites are involved than before and these include Fen Drayton GP, Cherry Hinton Cement Pits, Waterbeach GP, Ely beet factory and the Ouse Washes. One unusual record was a moulting female at Royston sewage farm from 25 June to 12 July 1978. Figure 17 shows the monthly distribution of records and it can be seen that the majority of the records occur at times of passage, particularly in the spring.

Earliest date: 5 August (1975 Ely beet factory)

Latest date: 29 May (1984 Cherry Hinton Cement Pits)

(These exclude the exceptional summer record at Royston.)

Velvet Scoter *Melanitta fusca*

A rare storm-blown vagrant.

Ten records, listed below:

1 Jenyns recorded a male shot near Haddenham, (no date). This bird was deposited in the Cambridge University Zoology Museum. [Lack]
2 A male and a female shot near Ely around 1880. [Lack]
3 One at Earith Washes on 15 February 1940.
4 One on the Earith-Sutton Washes, 20 January 1943.
5 One at Cambridge sewage farm, 4 November 1952.
6 Three which were visiting St Ives GP (Hunts) crossed into Cambs during flight on 12 January 1964.
7 A female on the River Ouse at Ely on 14 October 1973 was diving for freshwater mussels which it was seen to swallow whole.
8 An immature male at Fen Drayton GP on 16 February 1979.
9 An immature male, in poor condition, was picked up at Wisbech in late February 1979.
10 A female at Fen Drayton GP, 9 November 1980.

Goldeneye *Bucephala clangula*

Pre 1934

Jenyns listed three records and Evans stated that it was an occasional winter visitor. Lack reported that it was a fairly regular winter visitor to the Washes (presumably both Ouse and Nene) in very small numbers, and that elsewhere it was only very occasional.

1934-1969

In the middle period the number of records per annum was usually small, perhaps three or four, but there was an exceptional period from 1955 to 1958 with many records; and in 1961 the largest total for a single site (84 on the Ouse Washes) was recorded. In all other years the pattern was of small numbers at one or two sites. The annual maxima for the Ouse Washes from 1959 onward are displayed in Figure 18.

Present Status

A regular winter visitor.

Present from November to mid April in small numbers on the larger areas of open water. The largest regular gatherings recently have been at Fen Drayton GP where up to 60 have collected, preferring this site to the nearby Ouse Washes where annual maxima have declined from 32 in 1972 to one or two in the recent years (see Fig.18). Elsewhere it has been recorded regularly (usually 1-4) at Landbeach GP, Mepal GP, on the River Cam near Bottisham Lock, Milton GP, Roswell Pits/Ely beet factory, Waterbeach GP, Wicken Fen and along the River Nene.

Earliest date: 29 September (1984 Fen Drayton GP)

Latest date: 4 June (1958 Milton GP)

Figure 18. Goldeneye – annual winter maxima on the Ouse Washes 1959-85.

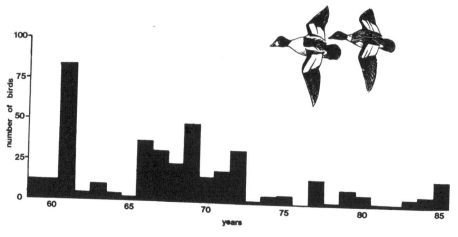

Smew *Mergus albellus*

Pre 1934

Evans described this species as 'a rare winter visitor', a summary with which Lack concurred, adding that they were more common in harsher winters. Four records from the nineteenth century are quoted by Lack, to which he added six in the twentieth. These were all of single birds, and where a month is mentioned it was always January. Most of these were of birds which had been shot.

1934-1969

There was a single record in the 1930s, and four in the 1940s, the most notable being during severe weather in January and February 1947 when up to 15 birds were noted on several occasions. It was in the 1950s that records became both more numerous and more regular. In this period the numbers were usually still small with perhaps one or two birds involved. The exceptions were in the years 1954, 1956 and 1963 when in common with other sawbills (see Goosander and Red-breasted Merganser) large numbers were recorded, for example 22 on the Nene Washes in 1954, 34-40 on the Ouse Washes in 1956 and 10-14 at the same site in 1963.

Present Status

Almost an annual winter visitor, usually on either of the Washes, in very small numbers.

Seen predominantly in January and February (90% of all recent records occurred in these months) at various sites but most particularly on the Ouse or Nene Washes. Numbers at any one site rarely exceed four although a total of nine were present on both the Ouse and the Nene Washes in February 1979, which proved to be another exceptional year with about 20-30 birds in the county, and in 1985 23 were counted along the Ouse Washes in January and 10 along the Nene Washes. Other sites which are visited are Fen Drayton, Landbeach, Milton, Mepal, Wimblington and Waterbeach GPs, the River Ouse at Ely, the River Cam and Wicken Fen.

Earliest date: 20 October (1971 Ely beet factory)

Latest date: 2 April (1956 Ouse Washes)

Red-Breasted Merganser   *Mergus serrator*

Pre 1934

Jenyns recorded one, from Swaffham (Bulbeck or Prior) 1840 and Evans described this species as 'a rare winter visitor'. Lack reported several further records: two from 1854, in the Wisbech Museum, from Whittlesey and Welney (was this a Norfolk record?); two mentioned by Saville (1854) and another by Newton (1847). He also recorded that Booth (1881) 'noted several individuals on the lodes and rivers around Cambridge in the winters of 1860 and 1861'. In the early part of the present century Farren received a specimen from Swavesey in October 1905, and one from Duxford in November 1923. A drake was seen at Cambridge sewage farm in November 1931. This makes a total of six records together with the influxes of 1860 and 1861.

1934-1969

Up to 1956 there were a further four records, birds at Earith in January 1939 (shot) and 1943, at Hildersham in 1949 and in 1954 on the Mepal part of the Ouse Washes. However, 1956 was a year of exceptional numbers, so much so that a record of 130 on the Ouse Washes was followed by 35 on the Nene Washes six days later. Subsequently records were of their more usual irregular nature: one in 1958, two in 1963, another in 1964 and one in 1967 during foggy weather in April.

Present Status

A rare and irregular winter visitor/spring migrant often forced inland by inclement weather.

This species remains an irregular visitor. The favoured sites are the area of the River Nene and its washlands, and the Ouse Washes or surrounding waters. On the Ouse Washes it has been recorded as follows: three in 1972 and annually from 1977 with the exception of 1980.

In the years 1979 and 1985 there have been minor influxes but in the other years the maximum number of records never exceeded three. In summary there have been no more than 30 records of this species apart from the years when there were large numbers present: 1860, 1861, 1956, 1979 and 1985.

Earliest date: 29 July (1972 Mepal GP)
Latest date: 30 April (1981 Ouse Washes)

Goosander   *Mergus merganser*

Pre 1934
Jenyns noted two records from Bottisham and Whittlesford around 1830 and there were two from Guyhirn dated 1840, in the Wisbech Museum. Evans described it as a 'rare winter visitor', and Lack quoted a number of records while describing it as an occasional winter visitor, regular on the Washes.

1934-1969
Goosanders were recorded regularly with one or two records per annum, principally on the Ouse or Nene Washes. In the 1950s there were some large gatherings, e.g. 47 on the Ouse Washes in March 1956, but in general there were less than ten birds at any one site at one time. There was an unusual record of three females at Over for about ten days in June 1963.

Present Status
An annual winter visitor from November to April at favoured sites, in small numbers.
The status has changed only marginally since Lack's time. This species is found on both the Ouse and Nene Washes in most winters, usually in greater numbers at the latter, and recent maxima are: 17 on the Ouse Washes in 1970, and 37 on the Nene in 1979. It is far more common than the Red-breasted Merganser and there seems to be a regular population of 20-30 birds. In common with the other sawbills there are years when unprecedented numbers occur (almost always associated with bad weather), particularly 1941/42, 1946/47, 1953/54, 1955/56 and 1979 when well over 50 birds were present at one time. In January 1985 a county total of around 100 (or more!) birds was probably the highest recorded. Away from the Washes the sites where Goosanders are recorded regularly are Fen Drayton, Landbeach, Milton, Mepal and Waterbeach GPs and Wicken Fen. Since oversummering has occurred on the Ouse Washes (of no doubt injured birds) no early or late dates are included.

Ruddy Duck   *Oxyura jamaicensis*
A feral species with some escaped birds.
All recent records are listed below:
One at Fen Drayton GP on 10 August 1975.

One flying down the River Cam at Horningsea, 13 May 1977.

A male on the Ouse Washes on 12 February 1978.

A male on the Ouse Washes during February 1979 was doubtless part of a nationwide surge outwards from the normal midlands stronghold due to harsh wintry conditions (Vinicombe and Chandler 1982).

A female at Roswell Pits, Ely, 15 November 1980.

A female at Fen Drayton GP, 31 October to 4 November 1981.

A male at Fen Drayton GP, 31 January to 8 February 1982.

One at Fen Drayton GP, 29 August 1983.

One at Fen Drayton GP, 19 December 1983.

Two on the mere at Wicken Fen, 31 December 1983.

A female on the Ouse Washes, 5 January 1984.

Singles at Fen Drayton GP on three dates in 1984; 10 and 27 January and 27 August.

A female flying north over the Ouse at the 'Fish and Duck', 20 January 1985.

On the Ouse Washes in 1985: a female, 29 May, and two males from 31 May to 14 June.

One at Whittlesey West Pit, 5 January 1986.

One at Chatteris GP, 13 December 1986.

One at Fen Drayton GP in the winter of 1986/87.

Two pairs on the River at Great Shelford, 28 March 1987, after a gale.

VINICOMBE K.E. and CHANDLER R.J. Ruddy Ducks during winter 1978/79. *Brit. Birds* 75. 1982.

EAGLES, HAWKS AND FALCONS

The two important factors concerning these birds have been the use of the organo-chlorine pesticides in the late 1950s (which were severe on Sparrowhawks in Cambridgeshire and affected the wintering numbers of both Merlin and Peregrine) and the very effective protection of the rare birds of prey such as Kite and Osprey which has led to visits from wandering and passage birds.

Honey Buzzard   *Pernis apivorus*

An extremely rare visitor.

Five records, all listed below.

1 One shot at Bottisham in September 1826. [Jenyns, Lack]
2 One shot at Hildersham in October 1826. [Jenyns, Lack]

3 A female (nesting?) shot near Newmarket in May 1862. [Lack]
4 One killed near Caxton in November 1876. [Lack]
5 One heading south-west over the A45, Newmarket by-pass, on 18 June 1976.

[Black Kite *Milvus migrans*

One flying over Histon, 28 May 1987. This record has not been scrutinised by the British Birds Rarities committee at the time of writing.]

Red Kite *Milvus milvus*

A very rare vagrant.
Lack reports that it formerly bred in the county (Jenyns) although it was noted to be in decline by 1849.
Seven records, all listed below:
1 One, in the Cambridge University Zoology Museum, was collected at Histon in March 1844. [Lack]
[One circling south-west above Over in February 1964 remained probable but not certain as the observer was not a birdwatcher.]
The return to regular breeding in Wales and the subsequent expansion to over 40 pairs seems the most likely explanation for the recent spate of records.
2 One gliding to the south over Little Downham, 4 June 1970.
3 One at the Ouse Washes, 22 May 1972.
4 One flew north over Elm, near Wisbech, 4 October 1984.
5 One at Outwell, 9 November 1984.
6 One at Abington, 8 May, and Pampisford, 9 May 1986.
7 One at Great Widgham Wood, 24-26 November 1986.

White-Tailed Eagle *Haliaeetus albicilla*

An extremely rare vagrant. Two nineteenth century records, listed below:
1 One obtained at Stetchworth in March 1847 having previously been seen in Dullingham. [Lack]
2 One shot near Chatteris in October 1875 was wrongly identified as a Golden Eagle in the first instance. [Lack]

Marsh Harrier   *Circus aeruginosus*

Pre 1934

Jenyns described this species as common in the fens and low meadows and it seems to have been breeding up to the period 1860-80, but was unusual thereafter. Evans stated that it was no longer breeding and was unrecorded by the early twentieth century. Lack gave four records: two at Wicken in 1928, one near the Cambridge sewage farm in 1932, and the last on Burwell Fen also in 1932.

1934-1969

While one or two were noted between 1934 and 1950 it was after the latter date that records became annual. Never were more than two birds involved, but sites such as Wicken Fen, Ouse Washes and Fulbourn Fen were visited regularly. There were no seasonal exceptions but more records occurred during autumn passage than at other times.

Present Status

A regular visitor, usually in spring, summer or autumn.
Found on the Ouse Washes, usually from April to October with one or two birds oversummering. In autumn (August/September) a roost often develops with a maximum of nine birds (1980) counted. At Wicken Fen individuals are frequently seen in April/May or August/September and although birds have been recorded in all months of the year winter records have become very unusual. Other favourite sites include Fowlmere watercress beds, Fulbourn Fen and the Nene Washes.

Breeding Status

Following increased summering by more and more birds the first breeding record this century occurred in 1981 with others in 1985, 1986 and 1987. It is quite possible that this species will establish itself as a regular breeding species in the future as it has in neighbouring counties.

Ringing Results

A young bird ringed in the county in June 1985 was found dead the following October at Nouakchott, MAURETANIA.

Hen Harrier *Circus cyaneus*

Pre 1934
Jenyns described this species as not uncommon and breeding on the ground; a statement which Lack doubts, suggesting confusion between this species and the Montagu's Harrier especially since there are no subsequent such records. Evans stated that it was 'not recorded recently' and was not breeding. Lack recorded it as a regular winter visitor to Wicken Fen.

1934-1969
Most records in this period were sightings of single birds in winter usually at, or near, Wicken or Fulbourn Fens, Fowlmere watercress beds or on the Ouse Washes. In some years, particularly the late 1960s, one or two birds stayed for longer periods during, and occasionally throughout, the winter period (November-March).

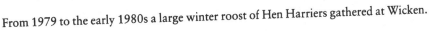

From 1979 to the early 1980s a large winter roost of Hen Harriers gathered at Wicken.

Present Status

An increasing winter visitor, some recent oversummering.

A very dramatic change has taken place over the last six or seven years. Up to 1977 the pattern of records was, as previously, rather sporadic with one or two birds seen irregularly through the winter at certain favoured sites (see above). In November/December 1978 the numbers on the Ouse Washes were so large that a roost formed holding a maximum of seven birds (3 males and 4 females/immatures), this being a local reflection of a national occurrence (Davenport 1982). In the same year birds at other sites became more permanent visitors. In the early part of 1979 a roost of up to 11 birds built up at Wicken with at least five more on the Ouse Washes, and by November/December 17 birds were roosting on the Washes and up to three on the Nene Washes giving a total in the county of around 25 birds. This pattern has continued and these roosts are regularly monitored (Clarke 1985). Roosts exist at Wicken Fen and on the Ouse and Nene Washes, and including that at Woodwalton Fen (Hunts) 20-30 birds may be in the area at any one time. During the day these birds frequently wander in the neighbouring countryside leading to many other sightings. Generally birds are now present from mid October until late April although one young male oversummered in 1985 and 1986.

CLARKE Roger. Hen Harrier roost survey in Cambridgeshire. *Camb. Bird Club Report* 59. 1985.

DAVENPORT D.L. Hen Harrier, Long-eared Owl and Short-eared Owl in 1978-79. *Brit. Birds* 75. 1982.

Montagu's Harrier   *Circus pygargus*

Pre 1934

Jenyns reported it to be rare. Evans seems very imprecise suggesting that it neither bred nor occurred, however records show that it bred at Wicken Fen about 1890 and, after a break, again in 1904. Lack states that one or two pairs were breeding in the 1930s and that wandering individuals were reported.

1934-1969

In the 1940s the breeding pattern continued although records were not always authenticated, and up to 1959 at least one pair summered in the county leading to possible breeding in both 1955 and 1958. Since 1960, however, all records refer to passage birds and in some years there were no sightings at all. In general the status diminished to that below.

Present Status

An irregular spring and autumn passage migrant.
There have been just over 20 records since 1970, by no means annual,
with about half during May and half in the period July-September. Most
of these records come from an area south of a line from Royston/Mel-
bourn to the Ouse Washes/Newmarket.

Breeding Status

After regular breeding in the 1950s there were no further records until
1981 when a pair raised young in a cereal field.
Earliest date: 13 April (1982 Ouse Washes)
Latest date: 28 October (1941 Rat Hall)

Goshawk   *Accipiter gentilis*

A rare vagrant.
Sixteen records to date, all listed below:

1 Evans and Lack mention a young bird obtained in Cambridgeshire
   which is to be found in Saffron Walden Museum.
2 A first-year male was seen briefly at the Cambridge sewage farm on 26
   April 1952.
3 One at Mepal on 23 January 1955.
4 One at Fulbourn Fen, 9 March 1955.
5 One at Welches Dam (Ouse Washes) on 9 March 1955 resembled the
   North American race.
6 One on the Ouse Washes, 26 February, 14 and 19 March 1956.
7 One on the Nene Washes, 28 June and 25 July 1956.
8 One at Milton, 15 October 1956.
9 One seen flying over Wicken Fen on 15 February 1959.
10 One around Madingley on 8 October 1959 was thought to be in the
   area until the end of November.

[One on the Ouse Washes (Pymore) on 16 December 1962 was considered
probable due to the lack of certain diagnostic characteristics in the de-
scription]

11 One on the Ouse Washes, 4 April 1965.

[One wearing jesses seen in Whittlesford, January 1976.]

12 A female, in moult, found dead at Chippenham in December 1977.
13 One at Welches Dam (Ouse Washes), 7 November 1985.
14 Two at Fen Drayton GP, 9 February 1986.

15 One on the Nene Washes, 10 December 1986.
16 An immature male on the Ouse Washes, 19 October 1987.

Sparrowhawk   *Accipiter nisus*

Pre 1934

Evans stated that this species was 'no longer common', while Lack summarised its position as 'sparsely distributed'. However, he considered it to be found throughout the county, although it was less numerous than the Kestrel.

1934-1969

Recorded regularly in the 1930s, 1940s and 1950s. A roost of six was noted in 1945 on the Ouse Washes. The effect of toxic chemicals was first apparent in 1959 (see Peregrine Falcon) and by 1960 the only known breeding pair was at Over. In 1963 there was not a single record, representing a most dramatic decline, which only showed signs of improving in 1967 when breeding was suspected and the number of records began to increase.

Present Status

A relatively scarce resident, more commonly seen in winter.
In the years 1983-87 the Sparrowhawk has been recorded with increasing regularity across the county, mainly out of the breeding season. At some sites, Wicken Fen for example, both male and female birds have been seen and there can be no doubt that records have increased since the early 1970s when there were only two or three per annum. In 1985 and 1986 the number of areas in which there were regular sightings suggests that there is now a spreading resident population. In neighbouring counties there are signs of recovery but it has been slow in Cambridgeshire.

Breeding Status

After the crash in the early 1960s there were a number of records of suspected breeding, particularly in the early 1970s, but the last proven of that period was as long ago as 1967. It took nearly 20 years for this species to overcome the effects of the toxic chemical disaster. In 1985 a pair raised a brood in the west of the county and in 1986 at least two pairs were successful.

Buzzard   *Buteo buteo*

Pre 1934

Evans described it as a 'very rare straggler', while Lack stated that it was an occasional winter visitor. Lack also pointed out that although Jenyns referred to breeding records it seems likely that he meant the 'Moor Buzzard' which was another name for the Marsh Harrier.

1934-1969

In the early years (1930s) it was not unusual for two birds to be present together, especially at Wicken Fen, but the post-war records are nearly all of single birds. The area around Wicken Fen is probably the most frequently mentioned but quite clearly birds were seen all around the county. Long stay was another feature of the 1930s records, but this also diminished after the war.

Present Status

An almost annual visitor.
Recorded in every month of the year except June; however, most records fall in the period August-November (Fig.19). Birds are seen all around the county, since the majority of sightings are of birds soaring on passage some are seen more than once as they move through. Single birds are

Figure 19. Buzzard – monthly distribution of all records.

normally involved, and there are usually only two or three records per annum, although in 1981, exceptionally, there were seven.

Rough-Legged Buzzard   *Buteo lagopus*

Pre 1934

Evans described this species as 'a very rare straggler'. Since by the time that Lack wrote his book there were only eight authenticated records the description was highly accurate.

1934-1969

In the period up to 1950 there were only two positive records; two birds were seen at Wicken Fen over the winter 1938/39 and one was found dead at Duxford in the winter of 1947/48. Two records in 1952 probably refer to the same bird which remained into 1953. One was seen at Wicken Fen in March 1955 and another, from November 1956, was seen near Balsham. There were no further records in the 1950s. One on the Ouse Washes in December 1962 was followed by one at Wicken Fen in March 1963 and either the same bird or another was seen on 6 April, travelling ENE over Cambridge sewage farm the following day. There were then three records in 1966, two in 1967 (when exceptional numbers were reported nationally and particularly in south-eastern England (Scott 1968), and a further two in 1968.

Present Status

A rare and irregular winter visitor.
There have been 11 records in the period 1970-85, some of which may

Figure 20. Rough-legged Buzzard – monthly distribution of all records.

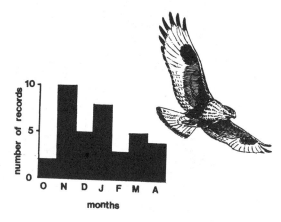

well have referred to the same bird. The present grand total is around 35. There were no records in ten of the years in this period. This species can be recorded at any location within the county and most sightings refer to single birds flying over. The monthly distribution of all the records is illustrated in Figure 20. Complications sometimes arise with identification when birds are either soaring high, or are seen only fleetingly, and there are a number of records which refer to Buzzard spp. which have not been considered for inclusion.

SCOTT R.E. Rough-legged Buzzards in Britain in the winter of 1966/1967. *Brit. Birds* 61. 1968.

Osprey   *Pandion haliaetus*

Pre 1934
Evans described this species as 'an exceptional winter visitor' which was something of an understatement since Lack found only three records as listed below:
One shot in 1839 near Wisbech now found in the Museum, and another in 1882.
One shot at Barrington in 1924.

1934-1969
A single bird spent about two weeks in September 1947 along the River Ouse on the Huntingdonshire/Cambridgeshire border. In 1952, when Ospreys were said to be 'plentiful' in south-east England, there were records in May at Cambridge sewage farm, and at Over.
One was seen at Cambridge sewage farm in September 1960 and another in May 1962.
A single bird was seen at Cherry Hinton in June 1962.
There was one at Wicken Fen in April 1965.
One on the Ouse Washes in May 1967.
One flew into Cambridgeshire from Huntingdonshire in May 1968 and two were reported in 1969.

Present Status
A rare spring and autumn passage migrant.
In the period 1970-87 there have been 23 records, of which 17 have been in the 1980s, showing a considerable recent increase which is only to be expected considering the success of the Scottish breeding population.

Since 1950 there has been a 19:10 ratio of spring to autumn records but those most recently seen are more evenly distributed at 14:9. Only once has more than one bird been seen when two were observed soaring over the Gog Magog Hills in August 1973.

Earliest date: 29 April (1965 Wicken Fen)
Latest date: 3 October (1983 Wicken)

Kestrel   *Falco tinnunculus*

Pre 1934

Evans considered this species 'not common'. Lack stated that a study by G.W.S. Thompson estimated the county breeding population in the 1930s to be around 300 pairs. Numbers were said to decrease in winter to about half. He also noted that it had bred on the ground at Wicken Fen.

1934-1969

This species remained much as Lack suggested for a time until the early 1960s when there was a clear sign of decline and the 1961 *Camb. Bird Club Report* raised the possibility of toxic chemical poisoning. In 1962 the decline was said to be not as bad as was feared, yet breeding was recorded in only 11 parishes. Throughout the next five years there were about 8-12 pairs noted per annum but the number of sightings, particularly in winter, was quite high. By 1969 the situation was just beginning to improve.

Present Status

The commonest diurnal raptor, widespread breeding species, passage migrant and winter visitor.

There has been a marked increase in both the breeding population and the number of wintering birds. In contrast to Lack's time this species increases in number in winter and counts along the Ouse Washes regularly exceed 20 birds in the autumn (September-December) with a maximum of 36 in September 1970. With the new areas of motorway/by-pass in the county birds can be seen regularly to a density of 2-3 per mile in places.

Breeding Status

By 1972 the breeding numbers had risen to around 15 pairs and in 1979 there were pairs present in the breeding season in 25 parishes. In 1983 the *Camb. Bird Club Report* suggested a county population of 100 pairs which is still a considerable reduction on the 1930 level of 300.

Ringing Results

A bird ringed in a nest at Madingley in May 1938 was found nearby in Coton in May 1945 showing that some birds stay in their natal area. One ringed fully grown at Spurn Bird Observatory in August 1951 was found at Wisbech two months later. (A migrant?) A bird ringed as a nestling at Sedburgh (Yorks), in 1966 was recovered in December of that year at Saxon Street (West Cambs). One ringed at Boxworth in June 1984 was found dead at Gedney (Lincs) two months later.

Red-Footed Falcon   *Falco vespertinus*

An extremely rare vagrant from eastern Europe.
Four records, listed below:

1 One in May 1909. [Lack]
2 One at Wicken Fen at dusk, 3 June 1971.
3 A female on the Ouse Washes, 15 May 1977.
4 An immature male at Fowlmere, 5 August 1987.

Merlin   *Falco columbarius*

Pre 1934

Evans described this species as a 'rare winter visitor'. Lack, however, stated that it was regular in winter, mainly in the fen districts such as Wicken and Fulbourn.

1934-1969

Not only did this species remain a regular winter visitor but over the war period and immediately post-war the number of records increased so that by the early 1950s there were quite good numbers in the fens each year and a roost at Wicken Fen contained up to five birds. This situation continued until in the late 1950s/early 1960s when the disaster of toxic chemical poisoning first reduced numbers and then, by 1964, eliminated this species as a wintering visitor altogether. In 1966 it was recorded once on each of the Ouse and Nene Washes. For the remainder of this period it was a most uncommon visitor with no more than two records per annum.

Present Status

A winter visitor recorded increasingly at present, having been decidedly uncommon for a period of around 15 years.

Up to 1977 there were generally one or two records per annum, usually of a single bird seen on the Ouse Washes. From 1978 on, the number of records increased and on the Ouse and Nene Washes and at Wicken Fen birds have been seen regularly during some winters in the 1980s, particularly 1983 and 1985. This suggests a return to the levels of the 1950s, indeed in 1985 birds were reported from 22 localities, mostly in fenland. This is strangely in contrast with the national breeding population which is in decline at present (Bibby and Nattrass 1986) and thus suggests that these birds have a Continental origin.

Earliest date: 19 August (1982 Ouse Washes)
Latest date: 2 May (1983 Nene Washes)

BIBBY C.J. and NATTRASS M. Breeding status of the Merlin in *Britain*. *Brit*. Birds 79. 1986.

Hobby   *Falco subbuteo*

Pre 1934

Evans described this species as a 'summer migrant that has bred with us'. Lack stated that one or two were reported on migration in August and September. He added that there had been about five breeding records including two as recent as 1932 and 1933 but considered that it may have been overlooked as a breeding species.

1934-1969

For the first part of the period the number of records was very small and

The Hobby is now an exciting addition to those species nesting regularly in the County.

usually referred to single birds seen, usually at Cambridge sewage farm, in the autumn. In the post-war period up to 1953 there were several years when no reports were received but from 1953 onwards the number of sightings increased, albeit slowly at first, and several records were received from all months of the summer until, by 1970, there were regular reports. In 1964 there were probably 1-2 pairs nesting and in 1965 a pair was present in a suitable location but in most other years there was little evidence of such activity. However, there is no evidence that the organochlorine incidence had any real effect on this species in Cambridgeshire since sightings increased over the same period that other raptors decreased.

Present Status

A summer visitor from late April to the end of September.
Recorded with increasing regularity in recent years and now a regular breeding species (see Breeding Status). Birds are seen throughout the summer and in many parts of the county although Hobbies are unusual in the fens away from Wicken Fen or the Ouse and Nene Washes. At these latter sites they are commonly seen in autumn specialising in feeding at the hirundine roosts which gather at that time. An interesting sighting was that of a bird observed flying near King's College on the Cambridge 'Backs' in July 1974.

Breeding Status

Sporadic breeding has been recorded since Jenyns' time and it seems likely that nesting in Cambridgeshire is a sign of a healthy national population. After the records of the 1930s and 1960s regular breeding has been reported since 1974 with a likely maximum of 12 pairs in 1987.
Earliest date: 31 March (1985 Wicken Fen)
Latest date: 18 October (1970 Hayley Wood)

Gyrfalcon *Falco rusticollis*

An extremely rare vagrant. One record only:
A single bird spent several weeks around Stretham in January 1940 and was independently seen by two observers.

Peregrine Falcon *Falco peregrinus*

Pre 1934
Jenyns stated that this species bred in Coton! Lack regards this statement

as doubtful with which I heartily agree. Evans considered it a 'rare straggler' and Lack described it as an occasional winter visitor and regular passage migrant seen mainly in the fenland region.

1934-1969

It was recorded more regularly in this period than either before or since. By the early 1950s it was frequently seen at preferred sites throughout the winter; this was generally in the fenland area around Ely and in 1942 a bird was seen living around the area of the cathedral. In 1959 the population crash due to the effect of toxic chemicals on both adult mortality and breeding success (Ratcliffe 1963, 1972) became evident with just a single record and the status changed to that of a most irregular visitor, with no sightings in five of the last ten years of this period.

Present Status

An unusual winter visitor, almost always in fenland.

At the time of writing it is possible that this species may be returning for longer wintering periods, particularly on or around the Ouse and Nene Washes, although some of the sightings have been obvious falconers' escapes. Up to 1981, however, records remained irregular in the extreme with no sightings reported in ten of the years 1970-85. In the 24-year period 1959-83 there were only 15 records; mostly single sightings of single birds. In 1984 one was present on the Ouse Washes in February, March and up to 12 April with possibly two on some dates. One was also present there 3-5 November. In 1985 among sporadic records on the Ouse Washes were two of a pair on 8 January and 17 March. On the Nene Washes one was seen on 6 July and 10 and 21 August. These are the only summer records in recent times. This species, though it remains scarce and irregular, is now recorded increasingly.

Earliest date: 6 July (1985 Nene Washes)
Latest date: 20 May (1982 Isleham)

RATCLIFFE D.A. The status of the Peregrine in Great Britain. *Bird Study* 10. 1963.
RATCLIFFE D.A. The Peregrine population of Great Britain in 1971. *Bird Study* 19. 1972.

GAMEBIRDS, RAILS AND BUSTARDS

Gamebirds are well nurtured within the county where shooting remains a common pastime. The changes in agricultural practices have however almost certainly had an effect on the population of the partridge species. The unusual rails have survived in small pockets such as at Fowlmere and

Wicken Fen and on the two larger washlands. Unfortunately, bustards are no longer recorded in the county.

[**Red Grouse**   *Lagopus scoticus*
A single record exists of a bird at Histon reported to Evans by Farren. No date is given.]

Red-Legged Partridge   *Alectoris rufa*

Pre 1934
Evans described this species as abundant and stated that it was first introduced into Suffolk in 1770, spreading to Cambridgeshire by 1821 if not earlier. Lack stated that by 1849 it had become plentiful, but added that at the time of his work it was not as common as the Grey Partridge.

1934-1969
Information was somewhat scarce which suggests that it remained a very common bird and by 1944 comment was made that in some parts of the county it exceeded the Grey Partridge in number.

Present Status
A widespread and common breeding resident.
Found on all areas of arable land and also recorded in suburban areas close to open fields. There continues to be much conjecture as to which of the two species is the more common although the Red-legged seems to predominate in the fenland area. It is worth noting that there has been some release of the closely related Chukar (A.chukar) in the county although records have not been collected.

Grey Partridge   *Perdix perdix*

Pre 1934
Evans described this species as 'most abundant' and Lack confirmed this by stating that it was abundant and widespread throughout the county.

1934-1969
An interesting record for 1947 describes birds landing on hedgerows in the fens in their bewilderment when the fields were flooded; otherwise the status appears to be unchanged.

Present Status

A common and widespread breeding resident.

There may have been some decline in numbers, yet despite reported problems associated with changes in agricultural practice (Potts 1970) it remains an easy species to locate around the county. It is undoubtedly more common on the chalk uplands than elsewhere and least common in the fenland area. The BTO common bird census results showed a 25% reduction in numbers in 1981.

POTTS G.R. Recent changes in the farmland fauna with special reference to the decline of the Grey Partridge. *Bird Study* 17. 1970.

Quail   *Coturnix coturnix*

Pre 1934

Jenyns noted that it was very plentiful but usually departed in September although some overwintering was recorded. Evans stated that it was 'not common' and seldom remained for the winter. Lack reported that nests were hard to find and considered that one or two bred on the chalk uplands. The last nest was found in 1909 at Cottenham.

1934-1969

Breeding records were hard to prove but there were plenty of instances of birds present and calling in due season, particularly in the Melbourn/Fowlmere, Gog Magog Hills, Newmarket Heath and Hildersham/Balsham areas. In most years after 1950 a minimum of 8-10 pairs were located. In 1963 there was an outstanding year when upward of 40 pairs were found scattered across both the extreme south of the county and in the southern fens (Burwell/Upware/Reach).

Present Status

A locally distributed summer resident.

Small population pockets remain in the traditional areas mentioned above and on the Ouse Washes. In the early 1970s there was a programme of introduction which led to increased records but this subsided quite rapidly and at present there are only one or two per annum.

Breeding Status

There is an understandable shortage of information since apart from calling these birds are very unobtrusive particularly when in tall crops.

Most records are of calling birds and at certain sites several birds are heard. The *Breeding Atlas* (Sharrock) shows this species present in four squares in the southern part of the county and proven breeding in the Ouse Washes square. In July 1980 an adult with a juvenile was seen on the Ouse Washes and this probably represents the most recent proven breeding record.

Pheasant   *Phasianus colchicus*

Pre 1934
Evans described this species as common and noted that most of those seen were of the ring-necked type. Lack stated that it was common and widespread and that it was a good fen bird and by no means confined to woodland.

1934-Present Status
A widespread and abundant resident.
The status of the Pheasant remains as stated by Lack. The high cost of maintaining a formal shoot has led to a more casual approach in some areas; nevertheless this does not seem to have adversely affected the population. A poor breeding season in 1981 led to cancellation of some shooting parties but generally numbers are maintained.

Water Rail   *Rallus aquaticus*

Pre 1934
Jenyns considered that it was not uncommon and Evans stated that he thought an occasional pair might still breed in the county. Lack noted that breeding was hard to prove but considered that pairs had been present at Chippenham Fen up to 1900, at Burwell in 1910 and at Cambridge sewage farm in 1932. Summering was recorded at Wicken Fen, and in winter this species was seen on the Ouse Washes, Fulbourn Fen, Cambridge sewage farm and Wicken Fen.

1934-1969
Recorded regularly at favoured haunts probably/possibly breeding at several but numbers were difficult to assess due to its retiring nature. At Fulbourn Fen, Fowlmere, Ouse Washes and Wicken Fen it was present throughout the year but at other sites, particularly the gravel pits, it was more of a visitor out of the breeding season. At Odsey regular breeding

was noted with a maximum of two pairs in 1960 and three broods in 1964.

Present Status

Resident, with a local breeding population, and a winter visitor. Its distribution is restricted to wetland sites with good vegetation cover (see above), but it is spasmodically recorded at less suitable areas but usually only in winter. Recent winter maxima are: Fowlmere, up to 15, Ouse Washes, up to 14 and Wicken Fen, up to 9.

Breeding Status

Breeding regularly takes place at Fowlmere (5-6 pairs), on the Ouse Washes (up to 14 pairs), and at Wicken Fen ('good numbers').

Spotted Crake   *Porzana porzana*

Pre 1934

Formerly bred, and considered by Evans to be common up to 1850. Booth (1881) found 2 or 3 nests at Wicken around 1860. [Lack] Lack quoted records of birds sent to Dr N.F. Ticehurst from Bourn in May 1896 and one from Cambridge in October 1893. Farren received specimens from near Cambridge in September 1897, September 1904, April 1914 and one from Wicken Fen in November 1904.

1934-1969

There were four positive sightings and one possible in this period, listed below:

One at Chesterton Fen, 31 October-14 November 1953.
One at Fulbourn Fen, 10-11 October 1957.
One at Fulbourn Fen, 13 March 1959.
[One at Cambridge sewage farm, 30 December 1961]
One was ringed at Wisbech sewage farm and when released flew across the River Nene into Cambridgeshire, 18 September 1966.

Present Status

An uncommon and highly local summer resident.
One or two dead birds have been found around the county but the population is mainly centred on the Ouse Washes and at Wicken Fen. Visual records are very rare but a single bird was ringed at Wicken Fen in August 1979.

Breeding Status

Since 1977/78 the large number of vocal contacts at Wicken Fen and on the Ouse Washes suggest that breeding takes place, with possibly 2-3 pairs at each site at present. Calling at a third location in 1983 might indicate an expanding population.

Little Crake   *Porzana parva*

An extremely rare vagrant.
Two records, listed below:

1 One obtained from near Barnwell, Cambridge, 16 March 1826. [Evans]
2 A male taken at Chesterton, Cambridge, in late March 1864. [Lack]

Baillon's Crake   *Porzana pusilla*

An extremely rare vagrant, which bred in 1858.
Three records, listed below:

1 One taken at Melbourn in January 1823 was the first British record. [Lack]
2/3 Two nests were found in the county, one in June and another in August 1858. These constitute the only British breeding records to date. [Lack]

Corncrake   *Crex crex*

Pre 1934

Evans stated that it was a 'common summer visitor' and Farren told Lack that it was a very common breeding species 50 years before he wrote, with a pair in every small grass meadow. By 1900 numbers had decreased but as late as 1925 there were six pairs between Waterbeach and Upware. Lack noted that it bred irregularly in 1934 and that, with some autumn migration, there were some birds still present in December.

1934-1969

Norris (1947), analysing the national decline, that was obvious even in the 1940s, stated that the main cause was the loss of haymeadows and he clearly foresaw the impending demise of this species. In Cambridgeshire it was recorded in most years up to 1957 when for the first time there were no sightings. In the period that followed there were only five records in 13 years and, whereas breeding had been recorded, albeit spasmodically, at one or two sites in the south of the county (Duxford/Ickleton, Babraham/ Gog Magog Hills and Hildersham/Balsham) it was no longer. The last such record was in 1955. Much of this loss seems to have been due to the gradual change of land use through the century and particularly the disappearance of meadowland. In this respect Cambridgeshire was merely a reflection of the national situation.

Present Status

An increasingly irregular and uncommon passage migrant.
This species is now simply a passage migrant in spring and autumn and is becoming quite rare. There have been only nine records in the years 1970-86, all of single birds and only one, at Girton in 1982, of any duration (17 June-5 July). In ten of these years none were reported. All this fits with the decline summarised by Cadbury (1980) which showed the population falling throughout western Europe.
Earliest date: 19 April (1960 Odsey)
Latest date: 4 October (1981 Fen Drayton GP)

CADBURY C.J. The status and habitats of the Corncrake in Britain 1978-1979. *Bird Study* 27. 1980.
NORRIS C.A. Report on the distribution and status of the Corncrake. *Brit. Birds* 40. 1947.

Moorhen   *Gallinula chloropus*

Pre 1934

Described by Evans as common and by Lack as common and widespread.

1934-1969

The only information of any interest concerns Cambridge sewage farm where in 1948 there was a maximum count of 140 in winter and 10 pairs bred. The following year the maximum count was 75. The status of this species remained the same throughout this middle period and apart from the severe effects of the 1962/63 winter when many corpses were found, breeding and distribution has been unchanged. In 1969 flooding on the Ouse Washes had a bad effect on breeding success and nests, eggs and corpses were all to be seen floating on the floodwater.

Present Status

An abundant and widespread resident, possible passage migrant and winter visitor.

Counts on the Ouse Washes suggest winter maxima of between 100 and 400. At other sites large numbers gather in hard weather, and in some arable areas gatherings of 10-30 can be seen feeding together often quite a long way from any water. This species is also commonly seen in suburban areas close to water.

Breeding Status

Common and widespread. Almost all streams and brooks have a resident pair and at suitable sites several pairs breed close together. The *Breeding Atlas* (Sharrock) shows every 10 km square occupied. Between 30 and 100 pairs nest on the Ouse Washes, and 38 pairs were counted on the Nene Washes in 1982.

Ringing Results

A bird ringed in October 1974 at Wicken Fen was found at Beaulieu Sur Layon, FRANCE, in November 1976. Another ringed at South Jaelland, DENMARK, in August 1962 was found dead at Fowlmere in December of the same year. These two give positive evidence of some immigration and emigration.

Coot *Fulica atra*

Pre 1934

Evans described this species as 'never common and breeds only in a few places'. Lack stated that it bred in deeper ponds and areas of open water and was very locally distributed, with some gatherings in winter.

1934-1969

Regularly recorded from all suitable sites throughout the county. As the gravel excavation increased so the population and distribution thereof rose. Early in the period there were gatherings of up to 100 at Cambridge sewage farm, but by the mid 1950s the largest numbers (200-400) were seen on the Ouse Washes. Like the Moorhen, it suffered considerably in the harsh conditions of the 1962/63 winter and many corpses were found along the Washes. By 1960 winter maxima on the Ouse Washes had reached up to 2000 and many of the pits had congregations of 100-200.

Up to 200 pairs (1969) nested on the Ouse Washes, while breeding in smaller numbers was recorded at most gravel pits.

Present Status

A common resident on all areas of open water.

In winter almost all the pits have in excess of 100 at peak times (usually February) and even small areas of open water have 1-2. Largest counts are on the Ouse Washes (Fig.21) and this site is the fifth most important for this species in Britain (Salmon 1987). At the larger pits the recent maxima are: Fen Drayton 1300 (1984), Waterbeach 576 (1983), Mepal 450 (1981) and Landbeach 355 (1983). On the Nene Washes a maximum of 530 was counted in 1981. During an average winter numbers tend to build up as the season progresses and the monthly counts on the Ouse Washes for the winter 1982/83 are presented in Figure 22.

Breeding Status

Small numbers breed on most pits and single pairs in the small areas of open water (e.g. Adams Road Bird Sanctuary, Cambridge). The largest numbers are found on the Ouse Washes where up to 200 pairs nest given suitable conditions. The *Breeding Atlas* (Sharrock) shows almost every 10 km square to be occupied.

Figure 21. Coot – annual winter maxima on the Ouse Washes 1965-85.

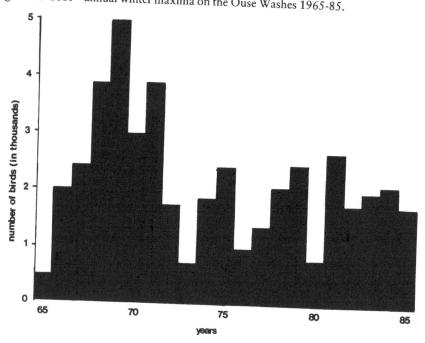

SALMON D.G. Wildfowl. In Salmon D.G. Prys-Jones R.P. and Kirby J.S. *Wildfowl and Wader Counts 1986-87* The Wildfowl Trust. 1987.

Great Bustard   *Otis tarda*

An extremely rare vagrant.
Jenyns wrote that formerly it was not uncommon on or around the heaths of Royston and Newmarket but by 1820 it was rare. Thackeray is quoted as saying that it bred at Fulbourn. Evans merely quoted Jenyns but added that there were attempts to reintroduce stock in nearby Norfolk.
The records that remain are listed below:
One reported to Jenyns, December 1826. [Lack]
One at Shelford in January 1831 is in the Cambridge University Zoology Museum. [Lack]
One from Littleport in January 1831 is also in the Cambridge University Zoology Museum. [Lack]
An egg was taken from a nest and presented to the University Philosophical Society in 1831, and a young bird was found at Trumpington. [Lack]
One at Caxton in December 1832. [Lack]

Figure 22. Coot – build-up of numbers on the Ouse Washes over the winter 1982/83.

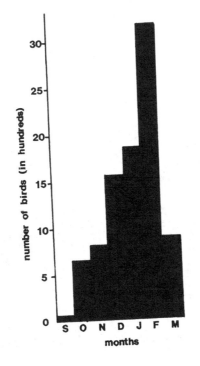

There is confusion concerning a record of a bird on Burwell Fen during February and March 1856 which was reported by Newton and recorded by both Evans and Lack. The original reports in *The Zoologist* however, indicate that two birds were present from the end of January, that one was shot and wounded (subsequent history unknown) and that only the un-harmed bird remained in the general region of Swaffham and Burwell Fens. Evans and Lack may have considered the original reports to be unreliable but there is no explanation given by either of them as to why these records are at variance.

J.M. Harrison received a stuffed female specimen killed at West Wick-ham on February 6 1880; this is not mentioned by Lack but appears in *The Zoologist* 1880 p.110 described by T.Travis.

[Birds were shot near Bottisham and Girton in 1902 but were almost certainly as a result of birds being released in Suffolk a year or two before.]

Crane   *Grus grus*

A rare vagrant.

Pre 1934

Evans gave an historical summary of this species starting in 1212 when King John obtained seven from Ashwell (which is just over the Hertford-

Great Bustards were a feature of the grassy heaths in the mid-eighteenth century.

shire border). William Turner (1544) stated that it bred in the fens although this does not necessarily mean Cambridgeshire. It seems to have occurred in large flocks even after 1678 and Ray noted that it was common in winter. The last historical record was of a bird killed near Cambridge around 1773.

1934-Present Status

Since the 1773 record there have been a maximum of 12 occurrences, (listed below): (These birds can occur in small invasions but it is very difficult to know on occasions whether the same or different birds are involved.)

1/2 In 1963 there was a national invasion when a total of 500 were recorded in Britain between 31 October and 3 November (Harber 1964). In Cambridgeshire one was seen flying over Fulbourn on 31 October, in addition 1-2 were seen at Wicken Fen from 8-11 November.

3/4 One on the Ouse Washes between 8 August and 27 September 1978, with another, or possibly the same, seen in the Whittlesey area.

5 An immature on the Nene Washes, 24-25 March 1979.

6 Two immatures on the Ouse Washes (only one in Cambridgeshire) 22-26 May 1979.

7 An adult at Wicken Fen, 18 May 1982.

8 Two on the Ouse Washes, 20 April 1984.

9 A first-year bird on the Nene Washes, 27 April-1 May 1984.

10 A sub-adult at Welches Dam (Ouse Washes), 20 April 1985.

11 One on the Ouse Washes from 28 April to 1 May 1985, definitely a different bird from the last.

12 An immature at Oxlode (Ouse Washes), 5 May 1986.

HARBER D.D. Influx of Cranes in October 1963. *Brit. Birds* 57. 1964.

Little Bustard   *Tetrax tetrax*

An extremely rare vagrant not recorded this century.

Four known records, listed below:

1 Bewick took his illustration from one obtained on Newmarket Heath. [Lack]

2 One in the Cambridge University Zoology Museum was shot at Caxton in December 1832. [Lack]

3 One from Welney Wash (presumably Cambridgeshire) in 1842 in Wisbech Museum. [Lack]

4 Another from Welney Wash in 1848 in the Wisbech Museum. [Lack].

WADERS

Of all the groups, waders have attracted the most interest over the years. Prior to Lack the value of the Cambridge sewage farm was only just being discovered and even post-Lack it was not until the 1950s that it realised its fullest potential for records. Sadly, for both birds and birdwatchers, the modernisation of this and other sewage farms (Peterborough, Royston and Wisbech) has drastically reduced their value as feeding grounds for this group of birds. An excellent three-part article by Graham Easy (1965, 1966, 1967) described the height and subsequent decline of the farm. Throughout this period, although much in the background, the Washes (Ouse and Nene) together with Ely beet factory attracted at times quite large numbers of interesting species. These sites are now the prominent wader recording areas in the county. Special mention must be made of Wisbech sewage farm where many rarities have been observed, for although it is just beyond the county boundary some of these species have been seen on the Cambridgeshire side of the River Nene and thus qualify for inclusion in this list and some early records came from the sewage farm itself since at that time it was part of the county. With some notable exceptions, Cambridgeshire is merely a temporary resting place for waders on migration; therefore although some wintering occurs, peak numbers are usually reached in the two periods April/May and August/September. For many regular migrants a histogram of the timing and numbers at Ely beet factory for the years 1978-80 has been included as an indication of the kind of passage that is recorded and particularly to show the annual variations that occur.

EASY G.M.S. Cambridge sewage farm. *Camb. Bird Club Reports* 40. 1966 (Part 1); 41. 1967 (Part 2); and 42. 1968 (Part 3).

Oystercatcher   *Haematopus ostralegus*
Pre 1934

Jenyns gave three records of single birds in the years 1827, 1835 and 1847. Evans described this species as a 'rare straggler'. There were five more records between 1929 and 1933, four of which were from Cambridge sewage farm. Late February/March records, as now, featured significantly among these early observations.

1934-1969

There were seven records between 1934 and 1952, since when Oyster-catchers were recorded annually although only two or three each year until the mid 1960s. Nearly all the records up to 1968 were of single birds, mostly in winter or on spring or autumn passage.

Present Status

Seen throughout the year, with a small breeding population.
With the increase in records, particularly in early spring, it was predictable that breeding would follow so the first nesting in 1971 came as no surprise. With the presence of breeding birds the number of sightings has increased and birds are now seen at places such as Ely beet factory, Fen Drayton GP, Waterbeach GP and Wicken Fen. Numbers away from the Ouse and Nene Washes remain small however, usually singles, and this species remains very rare away from the fenland north of Cambridge.

Breeding Status

In 1968 there were birds present on the Ouse Washes during May, and 1969 throughout the summer holding territory. In 1970 three pairs were displaying but no nesting resulted until 1971 when the first record was confirmed there and at the nearby Block Fen GP. This colony expanded rapidly to a maximum of 12 pairs (1977 and 1978) and by that time nesting had been discovered on the Nene Washes in addition, where up to 12 pairs have been counted, and on the Cam Washes where a single pair nested in 1980. Success has been varied but it seems that there is a potential total of 20-30 pairs on the washlands of the county, given suitable conditions.

Black-Winged Stilt   *Himantopus himantopus*

An extremely rare visitor.
Two records, both listed below:
[Howard Saunders gave a record for the county near Wisbech which Lack was unable to trace.]

1 During what seems to have been a minor national invasion by this species (see *Brit. Birds* 38. 1945) a pair visited Cambridge sewage farm, 2 May 1945.
2 Four on the Nene Washes in early April 1983, then a pair returned to breed in early May. Sadly, however, the clutch of four eggs was taken

and on the 17 June a pair (presumably the same) was seen briefly on the Ouse Washes before flying off south-west.

Avocet   *Recurvirostra avosetta*

A rare visitor.

Seventeen records to date, all listed below:

Lack mentioned that historically birds of this species were said to 'breed in the fens' but he noted that this does not necessarily mean in Cambridgeshire.

1 Jenyns mentioned one killed on the Cambridgeshire/Suffolk border in April 1828. [Lack]
2 An adult at Cambridge sewage farm in December 1936 remained until January 1937, this bird was first seen by Lack when passing in a train!
3 Two flew off from Cambridge sewage farm, gaining height and setting off south-east on 7 May 1947.
4 One fed among Black-headed Gulls at Cambridge sewage farm, 11-14 June 1947.
5 One at Peterborough sewage farm, 17-18 May 1948.
6 Six at Peterborough sewage farm, 23 June 1948.
7 One at Cambridge sewage farm, 26 May 1957.

In 1983, Blacked-winged Stilts attempted to breed on the Nene Washes.

8 One at Cambridge sewage farm, 25-27 September 1957.
9 Two flying north over the Ouse Washes at Pymore, 24 April 1966.
10 One at Ely beet factory, 12 June 1968.
11 Four at Ely beet factory, 3 June 1969.
12 One on the Ouse Washes, after a foggy night, 31 March 1974.
13 Two at Ely beet factory, 22-23 June 1975.
14 One on the Ouse Washes, 26-27 May 1981.
15 Twenty-five feeding on the Cam Washes at Upware on 27 March 1983 were seen to fly off eastwards.
16 One near Welches Dam (Ouse Washes), 6 September 1985.
17 Eight at Ely beet factory, 23 June 1986.

Stone Curlew   *Burhinus oedicnemus*

Pre 1934

Jenyns reported it 'breeding in plenty' while Evans described a decline. Lack stated that it was 'nowhere numerous but prior to cultivation was much commoner'. He also indicated that it bred from the eastern edge of the county up to the Gog Magog Hills and that Ennion saw quite large gatherings around Burwell and Fordham in September and October.

1934-1969

There are plenty of anecdotal records such as a bird seen walking along the railway line near Quy station! However, this species was not well documented until a review in 1952 based on fieldwork by A.E. Vine suggested a breeding population of around 70 pairs spread across the following areas: Chippenham 12, Newmarket Heath 10, Newmarket-Hildersham 30, Fulbourn 12 and South Cambridgeshire 13. In addition, post-breeding flocks of up to 120 birds were recorded. This situation seemed to remain true until the early 1960s when the first warning note appeared in 1963 and worried editorial comment appeared in the Camb. Bird Club Reports. In 1969 only 8 pairs were found, representing a most dramatic decline.

Present Status

An uncommon but regular summer visitor with a very local breeding population.
This species can be found in one or two pockets of the county and there are some quite large post-breeding flocks of up to 20 birds seen in some places.

Breeding Status

In the early 1970s up to 12 pairs were present in the county but between 1974 and 1977 breeding was not reported (although it was almost certainly taking place). In 1978 a positive attempt to monitor the population began and over the period to 1985 the county total has declined from 12 to 8 pairs (Jordan 1984). Jordan considered that this was due in the main to agricultural changes allied with increased disturbance and human-associated mortality (flying into wires and road deaths). Monitoring suggests, however, that this level, although low, is not unsustainable particularly since there are pairs just over the county boundary, so the population in the area as a whole is greater than suggested by the numbers in Cambridgeshire alone.

Earliest date: 19 March (1981 South Cambs)
Latest date: 8 November (1953 South Cambs)

JORDAN W.J. The Stone Curlew in Cambridgeshire. *Camb. Bird Club Report 58*. 1984.

Collared Pratincole   *Glareola pratincola*

An extremely rare vagrant.
Three records, all listed below:

1 One shot near Quy in May 1835. [Lack]
2 One (Pratincole sp.) on the Ouse Washes at Welches Dam on 7 June 1977.
3 One on the Ouse Washes at Purl's Bridge from 19 June to 3 July 1983 as the floodwater subsided leaving shallow pools.

Black-Winged Pratincole   *Glareola nordmanni*

An extremely rare vagrant.
One record only:

A single bird at Block Fen GP, 11-12 October 1982.

Little Ringed Plover   *Charadrius dubius*

1934-1969

Having bred in neighbouring Hertfordshire in 1938 (Parrinder and Parrinder 1969) it was no surprise when this species was first recorded in 1942 on 24 May at Cambridge sewage farm. Second and third records were similarly in May at the sewage farm in the years 1945 and 1948 this latter record marked the beginning of annual records. In 1949 the first autumn sighting was reported and by 1951 there were several records per annum. The first breeding record at Hauxton GP in 1952 was exactly ten years after the first sighting. There was then a rapid increase to around 4-5 pairs followed by a period of consolidation.

Present Status

A regular summer visitor with a small breeding population and a passage migrant.

Present from late March to late September, with a breeding population of between 10 and 30 pairs. Found at many wetland sites and particularly at Cambridge sewage farm, Ely beet factory, and the larger gravel pits. Passage numbers usually peak at between 10 and 20 (see Fig.23) at the two former sites.

Figure 23. Little Ringed Plover – annual autumn migration at Ely beet factory 1978-80.

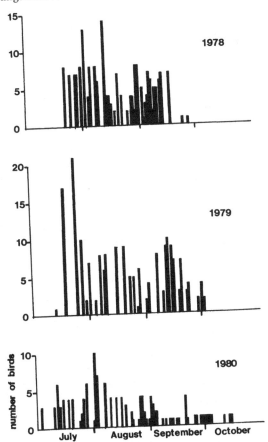

Breeding Status

Following the first record in 1952 breeding has been recorded annually, predominantly at the various gravel pits with some individual pits supporting more than one pair (up to 4 pairs at Waterbeach). Early records centred on the sewage farm population but rapid development meant that by 1957 there were 7-10 pairs in the county. However, while new sites were colonised some of the old sites became unsuitable, such as Wicken Fen where the mere became overgrown; thus this count was not exceeded until 1973. There was a steady build up in the 1970s linked with the development of the pit system so that in 1980 there were 27 pairs at 20 sites. Since then the situation has deteriorated with numbers declining alarmingly, and in 1985 only 5 pairs were reported.

Earliest date: 20 March (1983 Cam Washes)

Latest date: 29 October (1973 Cambridge sewage farm)

PARRINDER E.R. and PARRINDER E.D. Little Ringed Plover in Britain 1963-1967. Brit. Birds 62. 1969.

Ringed Plover   Charadrius hiaticula

Pre 1934

Evans described this species as an uncommon straggler. Lack reported that it was a regular summer visitor, breeding on the eastern border of the county, but more common in the Breckland. It was also a passage migrant both in spring and summer with numbers in excess of 100 recorded at the sewage farm in both seasons in 1930.

1934-1969

The general trend outlined by Lack continued, but the numbers quoted above proved quite exceptional, something between 20 and 40 were more normal maxima. Cambridge sewage farm was the main source of records and in 1952 a pair were present throughout the summer with breeding suspected but not proven. In 1955, however, a pair bred at Kennett GP, the first such record for a long time. With the demise of the sewage farm the number of records decreased.

Present Status

Recorded annually from February to October/November, with both a summering population and spring and autumn passage.

The favoured sites are the Ouse and Nene Washes (particularly in spring),

Ely beet factory and Wicken Fen (in autumn). The timing and extent of passage is variable as is shown in the histogram (Fig.24).

Breeding Status
Following the recent renewal of breeding in 1955 such records were for a while rather sporadic, usually based on the gravel pits and the sewage

Figure 24. Ringed Plover – annual autumn migration at Ely beet factory 1978-80.

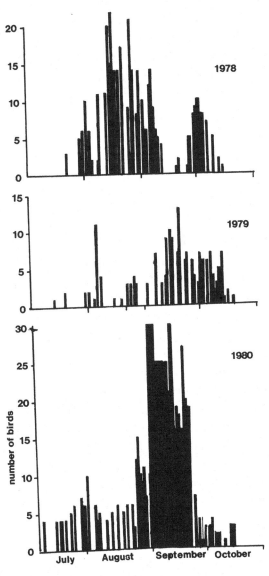

farm. In 1962 there were 5-6 pairs but in many years none were reported. In 1973, however, a new phase of regular breeding began with 4-10 pairs each year centred on the pits and Washes.

Kentish Plover   *Charadrius alexandrinus*

A very rare vagrant.

Seven records, all between 1943 and 1960.

1 One at Cambridge sewage farm in August 1943.
2 A female at Cambridge sewage farm, 23 August 1944.
3 A male at Cambridge sewage farm in the early morning of 22 August 1945.
4 One at Cambridge sewage farm, 31 August 1949.
5 Two at Cambridge sewage farm, in the early morning, 30 April 1950.
6 One at Cambridge sewage farm on 4 and 8 June 1951.
7 One on the Cambridgeshire/Norfolk border of the Ouse Washes on 18 April 1960.

Dotterel   *Charadrius morinellus*

Pre 1934

Jenyns gave the impression that even by 1849 this species was less numerous than formerly and records suggest that quite large numbers on migration met with their death to provide sport for the local inhabitants. Evans described it as a spring and autumn passage visitor but rare. Lack quoted only four records for the period 1900-34, in May 1905, in May 1906, in September 1929 and in October 1931.

1934-1969

Apart from a single record in August 1938 the first record was of 15 at a farm in the Heydon/Melbourn area in May 1949. Birds have been seen in this area during spring almost every year since. A report in 1950 that birds had been seen in the Balsham area led to further records and a suggestion based on information received that this had been a traditional site for many years. A 1951 report from Chatteris likewise inferred that birds were annual visitors. Numbers generally in this period were of the order of 1-20; exceptionally there were up to 60 in the Littleport area in May 1957 and 28 at the Heydon/Melbourn site in May 1962.

Present Status

An uncommon but regular spring, and rare autumn visitor.
Almost all the recent records occur in May, along the chalk line from Royston to Newmarket, although some birds move into the fens before flying north. Numbers are generally small (1-20) but during May 1980 a most exceptional passage took place (Easy 1980) when very large numbers (up to 187 in fenland) were recorded across the county. Although it was considered likely that a few birds were counted at more than one site this represents an unprecedented phenomenon. Autumn passage does occur, although it is rarely reported, and never involves the numbers that pass through in the spring.

Earliest date: 14 April (1981 Ouse Washes)

Latest date: 19 December (1961 Fulbourn)

EASY G.M.S. The spring migration of Dotterel. Camb. Bird Club Report 54. 1980.

Golden Plover   *Pluvialis apricaria*

Pre 1934

Evans wrote that it was not uncommon from autumn onwards often in large flocks. Lack described it as a common winter visitor from the first week of October to late March, occasionally seen in April. Flocks of up to

A remarkable Fenland Dotterel gathering in the spring of 1980

1000 were recorded, and some passage birds were noted at Cambridge sewage farm in May and during the August/September period.

1934-1969

In the early part of this period, up to 1950, there were annual records of winter flocks, generally between 50 and 500-600 strong, in areas close to Cambridge. The first large numbers were reported in the early 1950s when flocks of 1000 or more were found, usually in these same areas (Milton/Cambridge sewage farm), and on the Washes, both Ouse and Nene. Exceptionally there was a roost of 3500 near the sewage farm during hard weather in late January 1952. In the 1960s numbers seemed to be higher in spring than previously particularly in 1961 where the Washes amassed record numbers (Ouse 5500 and Nene 3500).

Present Status

A local but quite common winter visitor.

Records suggest quite a large number of fragmented flocks feeding in traditional areas throughout the county and this was confirmed by the BTO enquiry (Bircham 1978). Highest numbers at any one site are reached in spring (late March/April) with up to 1000 at some more favoured areas. Since the geographical proximity of several parties is so close it is often difficult to divide flocks but in general the following can be considered as the most traditional areas: Ouse Washes, Nene Washes, Fulbourn-Milton-Impington, Comberton-Toft, Trumpington-Grantchester, Newton-Thriplow-Fowlmere. Brinkley-Six Mile Bottom-West Wratting. Other flocks can be seen around the county and some passage takes place. Generally at the traditional sites numbers vary between 200-300 and 1000 with not all the birds in one block.

Earliest date: 10 July (1971 Ouse Washes)

Latest date: 10 May (1936 Cambridge sewage farm)

BIRCHAM P.M.M. BTO Golden Plover and Mute Swan Enquiry. *Camb. Bird Club Report 52.* 1978.

Grey Plover   *Pluvialis squatarola*

Pre 1934

Lack recorded two specimens in Wisbech Museum obtained in 1844 and 1845 at Guyhirn. He summarised its status as a regular passage migrant and quoted eight records between 1926 and 1933, half in spring and half in autumn.

1934-1969

With records mainly from Cambridge sewage farm this species emerged as a passage migrant more common in spring than autumn. Numbers were generally small (1-10), more often singles, and the duration of their visits were brief (usually 1-2 days). A proportion of the records were of birds seen or heard in flight over less suitable sites.

Present Status

A visitor mainly in winter and on passage.

In recent times this species has been recorded in every month of the year and once a warden was permanently present on the Ouse Washes (1969) it became clear that quite a few individuals pass through each spring (April/May) as well as occasional visitors at other times of the year. Most records refer to single birds but gatherings of up to ten have been seen. The situation is similar on the Nene Washes while the odd bird can be found at any of the well-known wader sites: Ely beet factory, Fen Drayton GP, Waterbeach GP and Wicken Fen.

Sociable Plover   *Chettusia gregaria*

An extremely rare vagrant. One record only:

A single bird was present with a flock of Lapwing in the fields between Horningsea and Fen Ditton from 18 to 30 October 1976. The local farmer thought that it had been present for about two weeks before it was identified. This was the fifteenth British record.

Lapwing   *Vanellus vanellus*

Pre 1934

Evans described this species as 'common in many places'. Lack reported that it was a widely distributed breeding species with some large flocks of several thousand out of the breeding season.

1934-Present Status

The most common and widely distributed species of this group of birds. Lack's description is still as accurate as when it was written. Breeding is widespread (see below) and the flocks are very common throughout the county as long as the land remains accessible. Considerable movements take place both within and through the county, and the largest of these follow two well-reported patterns. In late May the post-breeding adults, and by July all ages, migrate into and through the county. Also there is movement in to and out of the county before, during and after hard-weather periods. At this time birds move away south and west in front of the frost and snow and return following the thaw.

Breeding Status

Large numbers breed on the Ouse and Nene Washes with over 300 at the former and up to 150 at the latter. The breeding birds of wet meadows survey (Cadbury and Rooney 1982) showed a higher density on the Nene than on the Ouse. A further 26 pairs were reported from the Cam Washes, elsewhere the county has little pastureland but wherever conditions are appropriate a pair or two may be found.

Ringing Results

One ringed in Latvian USSR in June 1935 was found at Ely the following November. Another found at Ely in April 1947 had been ringed at Zuid in the NETHERLANDS, in October 1940. One ringed in Estonia, USSR, in June 1955 was recovered in Chatteris in March 1956. One ringed at Burwell, 15 May 1961 was recovered at Muron (Charente Maritime), FRANCE, 29 March 1962. Two ringed at Fowlmere in August 1974 and October 1976 found dead in Baranovichy and Moscow both in the USSR in April 1980 and September 1978 respectively.

CADBURY C.J. and ROONEY M.S. A survey of breeding Wildfowl and Waders of wet grasslands in Cambridgeshire in 1980 and 1982. *Camb.Bird Club Report 56. 1982.*

Knot   *Calidris canutus*

Pre 1934

Jenyns gave one record for 1826 but also claimed that it was 'previously to be met with near Ely in autumn'. Foster recorded four near Wisbech in late July 1851. [Lack] Evans described it as an occasional straggler usually towards winter. Lack's records concern Cambridge sewage farm and are as follows: September 1930, September and December 1931, and June 1933 when two adults in full summer plumage were present.

1934-1969

A relatively rare bird prior to Lack, this species has subsequently been more regularly recorded, mainly due in the first instance to the observations at Cambridge sewage farm where it was seen annually until the late

1950s, and subsequently on the Ouse Washes. Numbers: generally 1-6, exceptionally 15 at the latter site in February 1958.

Present Status

A spring and autumn passage visitor.
Recorded increasingly as a regular annual visitor more often on migration in either spring or autumn but sometimes in winter (see Fig.25), usually on either the Ouse or Nene Washes, but occasionally elsewhere; Ely beet factory for example. Numbers: usually single birds but exceptionally 25 around Fen Drayton GP in December 1980. The duration of stay is usually very short and may in some circumstances be associated with prevailing weather conditions.

Sanderling   *Calidris alba*

Pre 1934

Lack noted that this species was first recorded in 1929 and was seen quite often thereafter in spring and autumn always at Cambridge sewage farm. A record of 26 there in spring 1930 remains quite exceptional.

Figure 25. Knot – monthly distribution of all records.

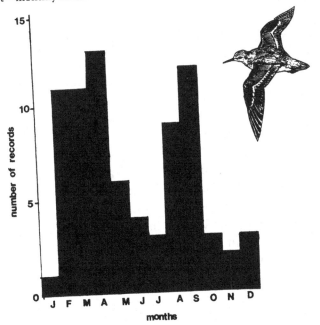

1934-1969

During the first part of this period this species remained a regular passage visitor almost exclusively at the sewage farm; the only other site where it was recorded was Peterborough sewage farm. Numbers: usually singles but sometimes 4-6, exceptionally 10 in May 1939. After 1957 there were single records in each of the years 1959, 1966 and 1968.

Present Status

An uncommon passage visitor.
Sanderling are seen mainly on the Ouse Washes or at Ely beet factory. Of the 11 records since 1969 nine have occurred in May (between 4th and 20th), the other two were in July and September. Numbers: usually singles, sometimes 2-4.

Little Stint   *Calidris minuta*

Pre 1934

Evans stated that the original British record for this species was in Cambridgeshire in 1849 but that it was not recorded subsequently, a statement which is countered by a record of one in May 1853 at Willingham quoted by Lack. Records at Cambridge sewage farm from the 1930s began a period of regular passage through the county.

1934-1969

At first it was seen only at the sewage farm, predominantly in autumn, with occasional spring passage. Numbers: usually 1-10, exceptionally 18-22 in 1957. Latterly (post-1960), records from Ely beet factory gradually began to assume major importance.

Present Status

A regular passage migrant, more common in autumn than in spring. Apart from Ely beet factory and the Ouse Washes, which are the preferred sites, Nene Washes, Waterbeach GP and Wicken Fen are all visited in some years. Numbers vary between 1 and 20 at the two popular sites and 1 and 10 at the others. The volume and timing of the autumn passage is extremely variable as can be seen in Figure 26.
Earliest date: 12 April (1981 Ouse Washes)
Latest date: 26 December (1967 Ely beet factory)

Temminck's Stint   *Calidris temminckii*

Pre 1934

Jenyns noted one from Foulmire (Fowlmere) and there are three records from Wisbech (1843, 1849 and 1854). [Lack and Evans] There were no more records until the 1930s at Cambridge sewage farm when in May and August 1930 and May 1931 one was recorded which Lack suggested was the same individual.

1934-1969

There were no further records up to 1938, but thereafter it became more frequent, for instance between 1948 and 1957 it was an annual visitor to Cambridge sewage farm in spring and/or autumn. Numbers: almost always single birds, rarely two and on one occasion up to five (August-September 1954). Spring records fell within the period 10 May-3 June,

Figure 26. Little Stint – annual autumn passage at Ely beet factory 1978-80.

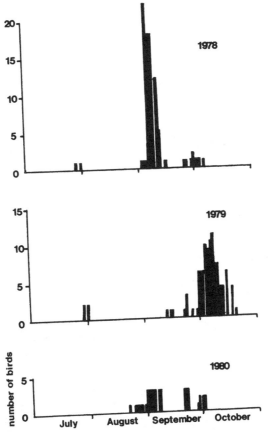

while in autumn birds could be found between 17 July and early September. In the period 1957-69, with the decline of the sewage farm, there were only three records.

Present Status
A rare passage migrant.
Irregular, seen in only nine of the sixteen years. Apart from one record of two birds in 1975 there were only single records per annum, of single birds. Marginally more of the records are in the May period than autumn, and only the Ouse Washes and Ely beet factory have been visited more than once.
Earliest date: 10 May (1950 and 1951 Cambridge sewage farm)
Latest date: 23 September (1968 Ely beet factory)

White-Rumped Sandpiper   *Calidris fuscicollis*

An extremely rare vagrant.
One record only:

A single bird flew into the county on 31 October from Wisbech sewage farm, having been caught and ringed on 24 October 1964.

Baird's Sandpiper   *Calidris bairdii*

An extremely rare vagrant.
One record only:

A single bird from Wisbech sewage farm that alighted briefly on the Cambridgeshire side of the River Nene on 29 July 1963.

Pectoral Sandpiper   *Calidris melanotos*
1934-1969

Unrecorded prior to 1948 when one was identified at Cambridge sewage farm and stayed from 29 May to 2 June, being the first British record of this species in spring. A second bird, seen at the same site the following October, was part of a remarkable series of observations of this species throughout Britain in that autumn (*Brit. Birds* 42. 1949).
There were seven further records in this period: in October 1952, September 1954, May 1956 and August 1957 at Cambridge sewage farm; and in the Septembers of 1960 and 1962 single birds from Wisbech sewage farm

flew into Cambridgeshire (although in 1962 two birds were present). In October 1966 the first record was received from Ely beet factory.

Present Status

A rare but increasingly recorded passage visitor, usually in autumn. There have been 14 records since 1969 of which eight have been at Ely beet factory. Almost all of these recent records involve single birds with the exception of two at Ely in 1985. There were three records in both 1971 and 1979. In the former year they were considered to be different individuals due to plumage differences; in the latter year birds were seen at different times at three separate sites and thus could have been the same individual. Some have stayed for up to a week, and one for a fortnight, but in general 1-2 days is the duration of stay. Out of the overall total of 22 records only two have been in spring.
Earliest date: 24 May (1956 Cambridge sewage farm)
Latest date: 22 October (1952 Cambridge sewage farm)

Curlew Sandpiper   *Calidris ferruginea*

Pre 1934

Evans stated that 'it has occurred on several occasions in the county' in autumn and mentions a record at Cambridge sewage farm in 1896. By Lack's time this species was recorded regularly on the sewage farm in autumn, but there were only four spring records.

1934-1969

The status remained unchanged and regular visits in autumn were noted almost exclusively at Cambridge sewage farm up to 1960 since when it occurred mostly at Ely beet factory with one or two records on the Ouse Washes. Numbers: small, generally 1-5, except in autumn 1954 at Peterborough sewage farm where a maximum of 65 were counted on 28 August and in 1969 at Ely beet factory where there was a maximum of 53 on 31 August. These records, discussed by Nisbet and Vine (1956) and Stanley and Minton (1972) seem to be associated with turbulent weather conditions during peak migration.

Present Status

A regular passage migrant predominantly recorded in autumn. Recorded annually in small numbers (1-10) at Ely beet factory and on the Ouse Washes. The main period of passage seems to vary from year to year within the months of August and September (see Fig.27). Spring records remain unusual with only seven between 1970 and 1984. Sites visited irregularly include Cam Washes, Fen Drayton GP, Royston sewage farm and Waterbeach GP.

Earliest date: 27 April (1975 Nene Washes)

Latest date: 6 November (1957 Cambridge sewage farm)

NISBET I.C.T. and VINE A.E. Migration of Little Stint, Curlew Sandpiper and Ruff through Great Britain in the autumn of 1953. *Brit. Birds* 49. 1956.

STANLEY P.I. and MINTON C.D.T. The unprecedented westward migration of Curlew Sandpiper in autumn 1969. *Brit. Birds* 65. 1972.

Purple Sandpiper   *Calidris maritima*

An extremely rare visitor.

One record only:

A single bird seen on Fulbourn Lode in November 1930.

Figure 27. Curlew Sandpiper – annual autumn migration at Ely beet factory 1978-80.

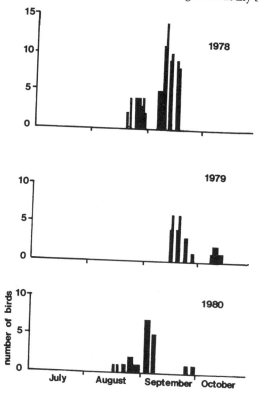

Dunlin   *Calidris alpina*

Pre 1934

Evans described this species as 'a rare straggler in autumn and winter'. Lack reported that by 1934 it was being recorded in each month, though not breeding, with some quite large congregations such as in November 1931 when there were over 100 at Cambridge sewage farm.

1934-1969

The status remained the same throughout this period with records coming at first from Cambridge sewage farm and latterly from the Ouse Washes. At the former site maxima of up to 50 were reported in autumn but at the latter a gradually increasing number, up to 600 in 1967, were seen in spring. From 1954 there were increasing numbers on the Nene Washes

Figure 28. Dunlin – annual autumn migration at Ely beet factory 1978-80.

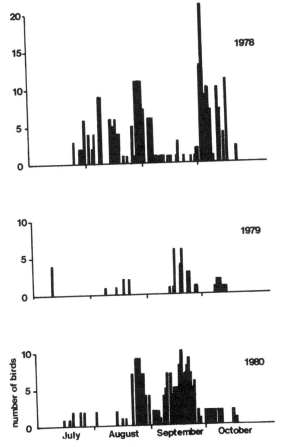

and by 1969 reports of up to 350 were received, again mainly in spring. At the time of the demise of the Cambridge sewage farm, Ely beet factory became more suitable and a similar autumn passage became apparent there.

Present Status

Recorded in every month of the year, but mainly out of the breeding season.

Favoured spring sites are the Ouse and Nene Washes with up to 860 (1971) at the former and 350 at the latter. In autumn the greatest numbers occur at Ely beet factory (see Fig.28) and Cambridge and Royston sewage farms. At many other sites, Cam Washes, Chesterton Fen, Fen Drayton GP, and Wicken Fen, numbers are small. Breeding has never been recorded.

Ringing Results

One ringed at Kattegar, DENMARK, 24 August 1966, was found dead at Wilbraham Fen, 3 January 1976.

[Broad-Billed Sandpiper   *Limicola falcinellus*

On 17-18 October 1946 a bird considered to be of this species was present at Cambridge sewage farm, although considered 'very probable' the identification was never confirmed.]

Stilt Sandpiper   *Micropalama himantopus*

An extremely rare accidental.
One record only:

A single bird at Wisbech sewage farm in July and August 1963 which flew over the county boundary.

Buff-Breasted Sandpiper   *Tryngites subrificollis*

An extremely rare accidental.
Three records, listed below:

1 One shot near Melbourn in September 1826 was the first record for Great Britain.
2 One at Peterborough sewage farm, 6-7 October 1953.
3 One at Peterborough sewage farm, 12-26 September 1962.

Ruff *Philomachus pugnax*

Pre 1934

Evans recorded it as a very common breeding species up to the beginning of the nineteenth century when it began to decline so that at the time of his list (1904) it was a visitor only. Lack concurred and added that Farren had told him that in the 1840s his father saw baskets full of Ruff and Reeves netted in the fens, for sale on Cambridge market. Evans stated that these birds were fattened up on bread and milk or on boiled wheat. Breeding ceased, according to Lack, around 1850. In April 1928 a party of 22 appeared on Fulbourn Fen where the males began their lek only to leave after a few days.

1934-1969

Immediately after Lack's publication (1934/35) there were records of one or two birds overwintering, said to be the first since 1896; and within two more years this species was present throughout the year at Cambridge sewage farm. This pattern continued throughout the 1940s and 1950s. At times numbers were quite high (up to 50), particularly during autumn passage, but generally 1-10 were recorded. A note in the 1950 *Camb. Bird Club Report* suggests that during winter they only visited the sewage farm in hard weather and were therefore wintering elswhere (on the Ouse Washes?). By the late 1950s and early 1960s birds were recorded at several sites in winter and spring, namely the Ouse, Nene and Cam Washes, Cambridge sewage farm and Ely beet factory. In 1962 the first breeding was reported (Cottier and Lea 1969, Easy 1968); however there

Part of a large influx of Ruff in the spring of 1965 on the Ouse Washes.

is a possibility that nesting had occurred previously but had been over-looked. By 1969 winter maxima of up to 200 were recorded.

Present Status

An uncommon but regular visitor and breeding species.

Present throughout the year on the Ouse Washes with a large build- up in late winter/early spring to maxima from 73 (1979) to 250 (1976). Else-where it is present out of the breeding season and mainly during periods of passage particularly in autumn. Ely beet factory, Cam and Nene Washes, Fen Drayton GP, Fulbourn Fen and Wicken Fen are the favoured sites. Numbers are very variable and on the washlands up to several hundred can be seen, elsewhere there are usually less than ten.

Figure 29. Ruff – annual autumn migration at Ely beet factory 1978-80 (scale twice other migration figures)

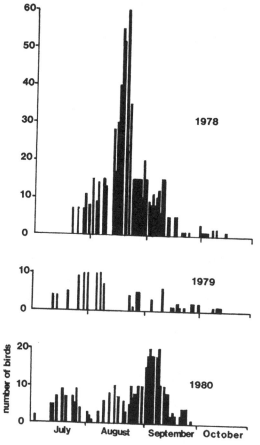

Breeding Status

Following the discovery of breeding in 1962, nesting has taken place in most years and lekking has been recorded annually. The tendency of the washes to flood in April/May in some years has drastically affected success, but generally around ten Reeves nest although 21 in 1971 was exceptional.

Ringing Results

Two birds ringed at Cambridge sewage farm in August 1956 and February 1957 were recovered in the USSR. The first was at Archangel in May 1957 and the second at Yakutsk (Soviet Asia) in May 1959.

COTTIER E.J. and LEA David. Black-tailed Godwits, Ruffs and Black Terns breeding on the Ouse Washes. *Brit. Birds* 62. 1969.
EASY G.M.S. The Ouse Washes. *Camb. Bird Club Report* 42. 1968.

Jack Snipe   *Lymnocryptes minimus*

Pre 1934

Evans described this species as 'fairly common'. Lack noted that it was nowhere numerous with parties of up to ten but usually singles seen from the end of October to March.

1934-1969

Apart from 1963 when it seems to have been unusually numerous in the country as a whole, and in Cambridgeshire in particular, the summary quoted above remained a very accurate assessment of its recorded status.

Present Status

A regular winter visitor.
Its status remains unchanged. Jack Snipe are recorded from the end of September to April at several sites including Fulbourn and Wicken Fens, and Landbeach, Milton and Waterbeach GPs. Numbers rarely exceed ten and one or two birds are usual. Since this species is unobtrusive it is quite likely that it is often present at sites which are not regularly visited and is therefore under-recorded.
Earliest date: 26 July (1935 Burwell)
Latest date: 21 May (1971)

Snipe   *Gallinago gallinago*

Pre 1934

Jenyns noted this species as plentiful after October, and Evans summarised its status as 'plentiful'. Lack recorded it as a common breeding species with some flocks on passage, particularly in autumn at Cambridge sewage farm.

1934-1969

No real difference from above. In 1953 a breeding estimate of 600-1000 pairs on the Ouse Washes was rather higher than at present, and the numbers at gatherings were rather higher, e.g. 700 on the Ouse Washes in 1959, and 1700 in January/February 1962.

Present Status

A common breeding species, winter visitor and passage migrant. Undoubtedly, apart from the Lapwing and possibly Golden Plover, the most numerous of all waders in the county in winter. Most open and wet areas, even woodland fringes attract visitors. At the more extensive sites large numbers gather and recent maxima are: Ouse Washes 3000 (1977 and 1979), Nene Washes 1000 (1985), Cam Washes 250 (1980) and Wicken Fen and Ely beet factory 100. On the fringes of Cambridge up to 100 can occur as at Chesterton Fen and 1-2 birds are seen in suitable areas in the city itself.

Breeding Status

The wet meadows breeding survey showed Cambridgeshire (including Hunts) to be one of the prime counties for Snipe (Cadbury and Rooney 1982). The Ouse Washes alone has anything up to 500 pairs and in 1981 400 were counted drumming on the Nene Washes. Many other sites have smaller numbers (e.g. Wicken Fen, Cam Washes, Fowlmere, Fulbourn/ Wilbraham Fen, etc.) and in many parts of the county the odd pair breed in hidden (wet) places.

Ringing Results

Two patterns of movement emerge from the results of ringing. Those birds ringed in Cambridgeshire in the summer, particularly as pulli, have been recovered in countries to the south such as a pullus ringed on the Nene Washes in June 1983 found dead in MOROCCO, and another ringed on the Ouse Washes in July 1974 shot at Finistere, FRANCE. A bird ringed at

Ely beet factory in late July, which may have been on passage, was recovered at Orense, SPAIN. Birds ringed, or recovered, in Cambridgeshire in winter, have origins in countries such as FINLAND, POLAND, SWEDEN and the USSR. One bird ringed at Wicken Fen in February 1961 was at St Dizant du Gua, FRANCE, two winters later in February 1963. Within Britain two birds ringed in late summer showed southward movement to winter in Dorset and Hampshire.

CADBURY C.J. and ROONEY M.S. A survey of Breeding Wildfowl and Waders of wet grasslands in Cambridgeshire in 1980 and 1982. *Camb. Bird Club Report* 56. 1982.

Great Snipe   *Gallinago media*

A very rare vagrant.
Six records, all listed below:
1 One at Bottisham in September 1839. [Jenyns, Lack]
2 One from Whittlesey is in the Saffron Walden Museum. [Lack]
3 One collected in the county in September 1898 taken to Farren. [Lack]
4 One collected in the county in October 1898 taken to Farren. [Lack]

[A bird was flushed from a path between a copse and ploughed land near Madingley in the winter in 1930. The observers felt sure it was of this species but the element of doubt was sufficient to record it as a probable.] [Lack]

5 One at Cambridge sewage farm, 16 October 1937.
6 One at Cambridge sewage farm, 23-24 August 1955.

[One was seen at Cambridge sewage farm on 12 October 1959 and although the description contained many of the features expected of this species it was rejected by the British Birds Rarities committee.]

Woodcock   *Scolopax rusticola*

Pre 1934
Neither Jenyns nor Evans knew of breeding in the county and both described this species as not common. Lack reported a breeding record at Chippenham Fen in April 1932.

1934-1969
A summary of status in the *Camb. Bird Club Report* for 1934 suggested a breeding population of 2-3 pairs at two sites (Chippenham Fen and Wimpole) and a comment that it did not occur in the Isle of Ely. The rather sparse records in the following period suggest either that it was not

common or that its rather shy nature led to few sightings. By the early 1950s regular breeding at Chippenham was noted as was wintering at six or seven sites. Into the 1960s breeding was recorded at other sites (Hayley Wood, Madingley and Wicken Fen) although the number of winter records remained small.

Present Status

A resident breeding species with some passage and winter visitors. Found at all the reasonably wooded wetland sites, particularly Wicken and Chippenham Fens, and is regularly recorded in the western part of the county (Hayley Wood, Hardwick Wood and the Gamlingay area). Rarely more than single birds are seen except in hard weather, such as in 1977 when ten birds were found in the gardens of Anglesey Abbey, while a report of 50 in the Gamlingay area in December 1981, with 32 shot nearby, must be quite unprecedented.

Breeding Status

Monitored by the number of 'roding' males, there is a strong population of 20 pairs or more at sites such as Wicken and Chippenham Fens (up to 7 pairs), Ditton Park Wood, Sawston Hall and Hayley Wood (2-4 pairs) and Whittlesford, Gamlingay, Dernford Fen, and Madingley (1-2 pairs). Many of these sites have only been recorded in recent years and thus although there appears to be an increasing population it is possible that this impression is due to increased observation.

Ringing Results

Two birds shot in the winter of 1977/78 at Hardwick and Arrington had been ringed in NORWAY and DENMARK respectively. Two birds ringed at Wicken Fen in August 1975 and 1978 were shot in Carmarthen (Wales) in December 1978 and 1979 suggesting either passage westward through Cambridgeshire or dispersal. A bird ringed as an adult at Wicken Fen in November 1979 was found in Sauerland, WEST GERMANY, in November 1982.

Black-Tailed Godwit   *Limosa limosa*

Pre 1934

This species formerly bred, but Jenyns stated that it was scarce and Evans wrote that the last nests were 'taken' about 1829. From this time until the emergence of Cambridge sewage farm there were probably no more than three records, at Cottenham in 1845 and at Wisbech in 1849 and 1850. Lack described it as an irregular passage migrant.

1934-1969

Recorded annually. Most of the records in this period were of birds on passage, more often in spring than autumn, and usually involved 1-2 birds at sites such as Cambridge and Peterborough sewage farms, or the Ouse or Nene Washes and Wicken Fen. The highlight of this period was the confirmed breeding record in 1952, and subsequently, on the Ouse Washes, after a period of suspected nesting. This was predicted by Morley (1939) who showed that with the increasing national occurrence a return to breeding was likely.

Present Status

Present in most months, a breeding species and passage migrant.

With the exception of the breeding stock this species is still predominantly a passage migrant, mainly in spring. Favoured sites include the Ouse and Nene Washes where spectacular numbers can occur (up to 250 at the former and 200 at the latter).

These large flocks usually contain a high proportion of birds of the

Figure 30. Black-tailed Godwit – the number of nests on the Ouse Washes 1952-86.

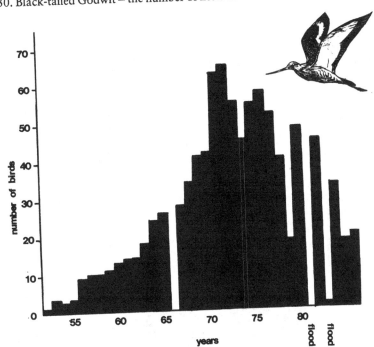

Icelandic race (*L.l.islandica*). Elsewhere regular records are received from Ely beet factory and Wicken Fen. During the high summer there are very few records away from the Washes and their immediate environs.

Breeding Status

Since first confirmed on the Ouse Washes in 1952 (Easy 1968, Cottier and Lea 1969) the number of pairs has risen from a single to a maximum of 65 in 1972 (see Fig.30). The success seems to have been high except in those years when unseasonal flooding has occurred. In 1985 three pairs nested at a second site.

COTTIER E.J. and LEA David. Black-tailed Godwits, Ruff and Black Tern breeding on the Ouse Washes. *Brit. Birds* 62. 1969.
EASY G.M.S. The Ouse Washes. *Camb. Bird Club Report* 42. 1968.
MORLEY A. The Black-tailed Godwit in the British Isles 1890-1937. *Brit. Birds* 33. 1939.

Bar-Tailed Godwit   *Limosa lapponica*

Pre 1934

Jenyns stated that this species was formerly abundant in the fens but there were only six firm records in the nineteenth century. Evans described it as 'an autumn and spring visitor, not common'. By Lack's time, however, the Cambridge sewage farm was beginning to attract irregular visitors.

Figure 31. Bar-tailed Godwit – monthly distribution of all records.

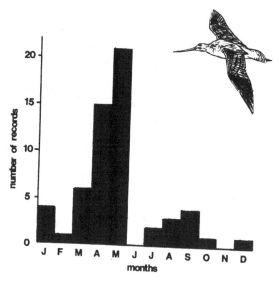

1934-1969

By no means annual. Most records in this period were of birds at Cambridge sewage farm or latterly the Ouse Washes, usually singles. Exceptionally 34 visited the former site on 4 September 1954 and large parties have been seen flying over the following sites: 65 over Milton on 31 August 1960, and 22 over Fulbourn Fen on 28 April 1962.

Present Status

A regular passage migrant.
Now more regularly recorded; annually since 1976. Numbers remain low, usually 1-5 but exceptionally 10-40, and the favoured sites remain the Ouse and Nene Washes, together with Ely beet factory. A seasonal bias in sightings has developed with about 90% in spring over the ten years up to 1986 (see Fig.31). Records in winter are extremely unusual, and summering has not been recorded.

Whimbrel   *Numenius phaeopus*

Pre 1934

Evans stated that this species was similar to the Curlew in status and thus was a rare autumn visitor, possibly more common in the north of the county than was supposed at the time. Lack described it as a regular passage migrant, more common in spring, most often recorded flying over, or heard at night.

1934-1969

Recorded every year from 1948 onwards, mainly in spring up to around 1960 since when the records have been equally distributed between the two migration seasons. Usually 1-5 birds, exceptionally 17 at Shippea Hill in 1950, and 18 and 22 at Littleport and Milton respectively in 1962. Some years were poor (1951, 1963 and 1968) and some were good (1956 and 1967).

Present Status

A regular spring and autumn passage migrant.
Passes through all parts of the county in both spring and autumn and birds are seen in flight as often as on the ground. Favoured sites include the Ouse and Nene Washes and Wicken Fen but many records are the result of casual observation by alert observers as birds fly over. Numbers:

normally small, rarely more than ten being seen together on the ground at any site. In the air, however, numbers can be higher, the most being 40 over Littleport in July 1970. The usual passage periods are mid April to mid May and mid July to, at latest, mid September.
Earliest date: 26 March (Ouse Washes 1980)
Latest date: 28 September (1954)

Curlew   *Numenius aquata*

Pre 1934

Jenyns stated that a nest was discovered by a Mr Fox! Evans described this species as 'apparently a rare visitor towards autumn; but it may be more common than is supposed towards the wash'. Lack reported single birds or small parties on migration and in winter, with birds often reported flying over.

1934-1969

Records continued to indicate spring and autumn passage with some limited winter visiting. Numbers were very small and records rather limited to the Ouse and Nene Washes and Cambridge sewage farm. Some were reported as 'heard flying over'. There was a report in 1950 of this species breeding on the Nene Washes and that it had done so since 1940 with a maximum of 20 pairs. There have been no such records since.

Present Status

Recorded in every month of the year.
Found at several wetland sites at any time other than during the breeding season. Nowhere is it more regular than on the Ouse and Nene Washes and birds are recorded at the former site in all months of the year. Numbers: usually 1-5 although up to 95 have been counted on or near the Ouse Washes. However, generally, as with the Whimbrel, the larger numbers are counted as birds fly over.

Upland Sandpiper   *Bartramia longicauda*

An extremely rare vagrant.
One record only:
A bird was shot about ten miles from Cambridge, 12 December 1854. [Lack].

[A possible was seen at Cambridge sewage farm in October 1932.]

Spotted Redshank   *Tringa erythropus*

Pre 1934

Pennant referred to this species as the 'Cambridge Godwit' which presumably means that it was regular in the fens. [Lack] There were four records in the nineteenth century, one in April 1838 near Ely, and three in autumn 1849 and 1851 from Wisbech and Ely. Evans stated that it was 'a rare autumn visitor usually singly but exceptionally in company'. Farren received one from Dullingham in August 1922 and by Lack's time it was a regular autumn visitor.

1934-1969

During the early part of this period this species was an unusual passage migrant recorded almost exclusively at Cambridge and Peterborough sewage farms, more often in autumn than spring. Numbers: usually 1-4 and staying for short periods only. In the mid 1960s records began to come from other sites and in some years exceptional numbers passed through (13 at Cambridge sewage farm in spring 1961 and 31 on the Ouse Washes in the autumn of 1968).

Present Status

Recorded in every month of the year, but mainly a regular visitor in both spring and autumn.
Found particularly on the Ouse Washes with regular counts of ten or more in autumn. Probably more widely reported than before. Elsewhere it has been noted at Ely beet factory, Fen Drayton GP and on the Nene Washes. Numbers in general seem slightly higher than in the past and where observations of single birds were the norm in the previous period 2-4 are now more common. Migration periods in general are late March to late May and early-mid July to late September. Recently there have been one or two records in December, January and February.

Redshank   *Tringa totanus*

Pre 1934

Jenyns noted this species as formerly plentiful but scarce by the time he was writing. Farren recorded it as common again by 1880s and Evans described it as 'not uncommon but local'. Lack stated that it was a common summer resident with large numbers passing through on spring and autumn migration, and some birds present in winter.

1934-1969

There was little change from Lack's summary, although it is possible that numbers were not quite as high as they are today. The Cambridge sewage farm was the focal point of passage birds with up to 100 present in late June/early July in the mid 1950s.

Present Status

Present throughout the year with a small breeding population. Found at all the wetland sites particularly on the Ouse, Nene and Cam Washes. Can be seen throughout the year but birds generally seem to move out following peak passage in July and return as the winter progresses.

Breeding Status

A summary in 1951 suggested a total 136 pairs in the county, yet two years later there were 100 on the Ouse Washes alone. Numbers fluctuate a little but generally around 100 pairs breed annually on the Ouse Washes (exceptionally 218 in 1977), and around 50 on the Nene. Elsewhere the Cam Washes, Chippenham, Chesterton, Fulbourn and Wicken Fens are used regularly by small numbers, and many other sites spasmodically. Cadbury and Rooney (1982) estimated that the county holds about 10% of all Redshank breeding in lowland England and Wales.

Ringing Results

A bird ringed at Seahouses, Northumberland, in August 1952 was found near Wisbech the following December. One ringed at Terrington in Norfolk in July 1974 was part of the breeding stock at Wicken Fen in April 1976. One ringed at Ely beet factory in July 1984 made the return journey to Terrington by August 1985. Another Wicken bird ringed in May 1976 was found dead in the Gironde district of FRANCE the following March.

CADBURY C.J. and ROONEY M.S. Survey of breeding wildfowl and waders of wet grasslands in Cambridgeshire in 1980 and 1982. *Camb. Bird Club Report* 56. 1982.

Greenshank   *Tringa nebularia*

Pre 1934

Evans described this species as an 'uncommon autumn visitor'. Lack stated that it was a regular autumn migrant to Cambridge sewage farm with occasional records from other parts of the county. He also noted some spring records.

1934-1969

As with other waders, the development of observations at Cambridge sewage farm led to increased information on this species and up to the middle 1950s this site, together with Peterborough sewage farm, provided the majority of the records in this period. Spring records usually consisted of 1-2 birds present for a single day, but numbers in autumn were higher, e.g. 24 in August 1952 and 30 in August 1954, and the duration of stay was longer. As the period progressed and the sewage farm declined, Ely beet factory and the Ouse Washes became the favoured sites.

Present Status

A spring, and more commonly, autumn passage migrant.
The favoured sites are the Ouse, Nene and Cam Washes, Ely beet factory, Cambridge and Royston sewage farms, Fen Drayton and Waterbeach GPs and Wicken Fen. Numbers: usually 1-4, but August maxima are between 6 and 12 at the most popular sites. Winter records are unusual but do occur thus making extreme dates information invalid, but generally passage takes place from mid April to mid or late May, and late June/early July to late September/early October.

Marsh Sandpiper   *Tringa stagnatilis*

An extremely rare vagrant.
Two records listed below:
1 One on the Nene Washes, 11 May 1981.
2 One at Purl's Bridge (Ouse Washes) from 30 April-8 May 1984.

Lesser Yellowlegs   *Tringa flavipes*

An extremely rare vagrant.
Three records listed below:
1 A single bird at Cambridge sewage farm from 29 March 1934 until at least 4 May from whence it moved to Burwell Fen where it was seen from 10 to 17 June. It then returned to Cambridge sewage farm from 23 July where it was frequently seen for more than a year up until 13 September 1935.
2 One at Peterborough sewage farm in May 1950.
3 An adult on the Cam Washes, 24-26 July 1983.

Green Sandpiper   *Tringa ochropus*

Pre 1934

Evans described this species as a rare spring and autumn visitor. By Lack's time, however, it was regular in autumn, although uncommon in spring, and some wintering had been noted.

1934-1969

Recorded annually, particularly at Cambridge sewage farm where birds were often present for long periods, and at other sites in smaller numbers. Autumn passage remained its most prominent period with maxima of 20-40 in peak years; but wintering was increasingly noted although numbers varied from year to year.

Present Status

An annual passage migrant and winter visitor.
There has been an increase in the number of records over the last few years. In addition, the number of wintering birds has risen to 5-6 at several

Figure 32. Green Sandpiper – annual autumn migration at Ely beet factory 1978-80.

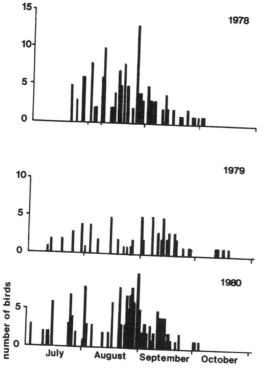

sites in some years. Spring passage remains light and summer records are extremely rare. Favoured sites are: Fowlmere watercress beds, Fulbourn Fen, Ely beet factory, Royston sewage farm and Waterbeach GP. This species is not recorded as frequently on the washlands as other waders and seems to prefer the smaller sites in the southern part of the county. Birds that stay for longer periods usually frequent dykes, ditches and streams. Autumn numbers at preferred sites peak at around 10-16 birds and Figure 32 shows passage at Ely beet factory in three autumns.

Ringing Results

Recoveries of two birds ringed at Fowlmere in August 1971 and 1972 were at Owthorpe (Notts) in late January 1979 and at St Omer (Calais), FRANCE, in July 1975 respectively.

Wood Sandpiper *Tringa glareola*

Pre 1934

Evans reported this species to be a rare autumn visitor. Lack described it as an autumn passage migrant at Cambridge sewage farm but not infrequently seen near other open waters; he said it was rare in spring.

1934-1969

Regularly seen at Cambridge sewage farm, more frequently in autumn, throughout this 35-year period, and in the later years at other sites such as Hauxton GP, Ely beet factory and the Ouse Washes. Numbers in this period were generally singles in spring, but 1-15 in autumn, especially at the sewage farm, and exceptionally 77 on the Ouse Washes in 1968.

Present Status

An annual passage migrant.
Recorded at all the favoured wader sites: Ouse, Nene and Cam Washes, Ely beet factory, Wicken Fen and Cambridge and Royston sewage farms. Numbers: usually 1-2, occasionally up to 4 but more than 4 is unusual. Seasonally more common in autumn (particularly since individuals seem to stay longer on site) than spring. There are annual fluctuations in the timing and strength of passage and in some years there are no spring records at all. Passage periods are late April to end of May and mid July to early/mid September.

Ringing Results

A bird ringed at Tetney Marsh (Lincs) on 10 August 1960 was found ten days later near March. This bird was possibly showing a southward migratory 'hedge hopping' movement.

Earliest date: 14 April (1974 Ouse Washes)

Latest date: 16 October (1968 and 1975 Ouse Washes)

Common Sandpiper   *Tringa hypoleucos*

Pre 1934

Evans described this species as a 'common summer visitor which may breed'. Lack, however, considered that it was mainly a passage migrant with some oversummering and overwintering. He recorded breeding as likely to have taken place between Bottisham and Upware in 1920 on information received from Farren Evans noted a bird near Byron's Pool,

Figure 33. Common Sandpiper – annual autumn migration at Ely beet factory 1978-80 (scale twice other migration figures).

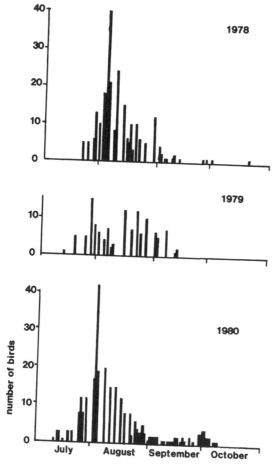

Trumpington, throughout the summer in several years around 1920 but no nest was found.

1934-1969

For the early part of this period records were mainly from Cambridge sewage farm but after its decline other sites of importance emerged such as Ely beet factory and Waterbeach GP. Generally passage was light in spring but far more considerable in autumn. Numbers in this period often peaked between 50 and 70 in August at Cambridge sewage farm but were much smaller at other sites.

Present Status

A regular passage migrant in spring and autumn.
Found at the usual favoured sites: Ely beet factory, Cambridge sewage farm, Waterbeach, Fen Drayton and Milton GPs and Fowlmere water-cress beds, but birds can be seen at almost any suitable site at peak passage time. Numbers: usually 1-10 but higher at Ely beet factory (usually up to 20, exceptionally 42 in 1980). Figure 33 shows three typical autumn periods. Passage periods are late March to early June and late June to end of October. Overwintering is very unusual but a single bird was to be seen in the general area of the Ouse Washes/Fen Drayton GP in 1975-76.

Ringing Results

Ringing at Ely beet factory has revealed that there is considerable site fidelity among the birds that pass through in autumn. Three birds caught in 1984 had all been ringed at that site in 1981 and the seasonal date of their recapture was within a week of the date of their initial ringing (Milwright 1984).

MILWRIGHT R.D.P. Report on bird ringing in Cambridgeshire. *Camb. Bird Club Report* 58. 1984.

Turnstone   *Arenaria interpres*

Pre 1934

A single record from May 1849 preceded the annual records at Cambridge sewage farm reported by Lack from 1927 onwards, usually in May or late July/August.

1934-1969

The pattern described above continued but, as with many other waders, by the late 1950s Cambridge sewage farm was no longer the source of regular sightings and in the years 1956, 1958, 1959 and 1960 none were recorded. In the 1960s there were fewer records in general and none in some years.

Present Status

An unusual spring and autumn passage migrant.

Unusually this species is more common in spring than autumn, usually on the Ouse or Nene Washes, or at Fen Drayton GP. Numbers: single birds are most likely but small parties of up to five are recorded. Figure 34 shows the monthly distribution of records.

Earliest date: 16 March (1977 Duxford)
Latest date: 16 December (1975 Ouse Washes)

Wilson's Phalarope   *Phalaropus tricolor*

An extremely rare vagrant.

Figure 34. Turnstone – monthly distribution of all records.

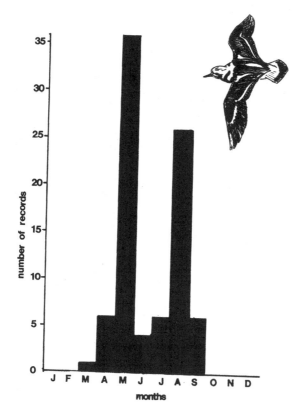

Three records listed below:

1 A single on the Ouse Washes from 3 May-4 August 1977.
2 A juvenile on the Nene Washes, 27-28 August 1979.
3 One stayed on the Ouse Washes for three weeks October/November 1984.

Red-Necked Phalarope   *Phalaropus lobatus*

A rare vagrant.

Thirteen records, listed below:
Lilford states that one was obtained at Whittlesey but the evidence of this record being within the county was considered by Lack to be inconclusive.

1 One obtained from the man who shot it at Cambridge sewage farm, 27 September 1896.
2 One obtained from Stapleford, October 1905. [Lack]
3 One at Peterborough sewage farm, 17 August 1952.
4 One at Cambridge sewage farm, 6-10 September 1953.
5 One at Cambridge sewage farm, 25 September-3 October 1957.
6 One on the Ouse Washes, 22-24 July 1973.
7 One on the Ouse Washes, 4 June 1975.
8 One at Ely beet factory, 27 September 1978.
9 One at Ely beet factory, 26-27 September 1981.
10 A female on the Ouse Washes (Purl's Bridge), 20-24 June 1983.
11 An adult in partial summer plumage was seen on the Ouse Washes from 28 to 31 July 1985.
12 An immature on the Ouse Washes (Purl's Bridge) from 30 August to 20 September 1986.
13 An immature at Ely beet factory, 31 July to 2 August 1987.

Grey Phalarope   *Phalaropus fulicarius*

A rare vagrant. Twenty records, all listed below.

1/2/3 Jenyns reported three shot in the fens 1819-20.
4 One in the Cambridge University Zoology Museum from Melbourn, date unknown. [Lack]
5 One at Barton, October 1854. [Lack]
6 A male at Swaffham Bulbeck in autumn 1866. [Evans]
7 Two shot near Cambridge about 1890. [Evans]
8 One at Littleport in November 1894. [Lack]
9 One at Ely, November 1904. [Lack]

10 One at Cambridge, 1904. [Lack]
11 One at Cambridge sewage farm, December 1912. [Lack]
12 One at Cambridge sewage farm, November 1929. [Lack]
13 One at Cambridge sewage farm, September 1930. [Lack]
14 One at Cambridge sewage farm, September 1931. [Lack]
15 An immature at Waterbeach GP, 25 November 1945.

[One at Cambridge sewage farm, 15 November 1946, was considered only a probable despite a perfect description being filed.]

16 One at Peterborough sewage farm, 16 May 1948.
17 One at Cambridge sewage farm, 8-20 November 1953.
18 One on the Ouse Washes (Purl's Bridge), 15 and 25 October 1981.
19 One on the Ouse Washes, 21 September 1984.
20 One at Ely beet factory in the aftermath of the hurricane, 16 October 1987.

SKUAS, GULLS, TERNS AND AUKS

Skuas, for obvious reasons, are very unusual visitors, the Arctic being the only species recorded with any regularity. As with other seabirds, sightings generally coincide with storm conditions on the coast.

Gulls vary according to the species involved. The habit of feeding at rubbish tips has brought increased records of some of the more unusual species, but those that are truly marine occur only in the kind of conditions described above. Work by the Cambridgeshire Gull Group, cannon netting at various rubbish tips, is beginning to show the patterns of local movement and seasonal migration of this family of birds. Their results to date are presented where relevant.

Terns are generally well recorded. Records of marsh species have always attracted attention but with the development of the gravel pits across the county, reports of sea terns have increased.

Auks, like skuas, are only seen during adverse conditions on the coast. The records all occur during the period September to March and on occasions many birds are compelled to come to land; these events being appropriately called 'wrecks'.

Pomarine Skua   *Stercorarius pomarinus*

An extremely rare vagrant.

Four records, listed below:

1 Jenyns reported one killed in the county near Cambridge in 1826.
  [Lack]

[Another reported by Jenyns at Tydd St Mary was probably not in Cambridgeshire.]

2 One on the Ouse Washes (at Oxlode) in November 1944.

3 Three immatures circling and slowly heading south at Milton, 4 December 1964.

4 One flying SSW over Mepal Washes (Ouse), 26 December 1985.

Arctic Skua   *Stercorarius parasiticus*

Pre 1934

Lack noted three records as follows:

One noted by Jenyns killed at Wisbech in November 1841.
One near Cottenham, 19 November 1861.
One from near Linton sent to Farren in September 1916.

1934-1969

There were eight further records in this period, mostly in September and all of birds noted flying over or found dead. The most interesting was of 16 (Skua species) seen flying over Milton on 20 September 1960 of which at least two were Arctic, and the remainder were probably the same, or possibly Long-tailed. Numbers: generally singles but one record of nine and another of seven.

Present Status

An uncommon storm-blown visitor.
A further 13 records almost all of which are of birds flying over. Numbers: between one and nine. Of all the 24 records only one, on 6 June 1966, has not been between 3 August and 15 December, showing this species to be almost entirely an autumn visitor. Where the race is mentioned most are pale-phase birds and when aged are as often adults as immatures.

[Long-Tailed Skua   *Stercorarius longicaudus*

One at Littleport, 26 September 1963, but the identification by its very long tail streamers seemed insufficient.]

Great Skua   *Stercorarius skua*

An extremely rare vagrant.

Four records, listed below:

1 One caught at Cottenham in October 1857. [Lack]
2 One from Melbourn sent to Farren in September 1910. [Lack]
3 One seen several times chasing gulls at Burwell Fen in February 1937.
4 Four flying over Milton 'at a great height' in the evening of 12 September 1975.

Mediterranean Gull   *Larus melanocephalus*

A very rare vagrant.

Eight records, seven of which are from the years 1981-86, all listed below:

1 One in a large gull roost on the Ouse Washes, 14-15 December 1974.
2 One flying south-east across the A10 near Waterbeach GP, 6 March 1981.
3 A second-summer bird flying south-west along the Ouse Washes, 19 May 1981.
4 One on the Ouse Washes, 19 January 1984.
5 An immature on the Ouse Washes, 11 March 1984.

Four Great Skua flew southwest over Milton at a great height on 12 September.

6 An immature following the plough at Milton, 16 March 1984.
   (It is possible that this and record 5 involve the same individual.)
7 A first-winter bird in a gull roost at Purl's Bridge (Ouse Washes), 24
   January 1985.
8 A second-winter bird on the Ouse Washes, 26 December 1986.

Little Gull   *Larus minutus*

Pre 1934

Lack referred to an immature labelled 'Cambridgeshire' to be found in the
Cambridge University Zoology Museum, and both Lack and Evans mentioned one shot at Wisbech in January 1862.

1934-1969

Curiously, with only two previous records, this species was recorded quite
regularly in the period following Lack's publication, so that by 1950 there
had been 12 records. During the 1950s and 1960s it was still seen
reasonably regularly and a further 17 records brought the total to 29 by
the end of 1969. Fourteen of the first 21 were at Cambridge sewage farm
and five of the last six were at the Ouse Washes showing how important
these two sites have been. Apart from five on the Ouse Washes in Novem-

Figure 35. Little Gull – monthly distribution of all records.

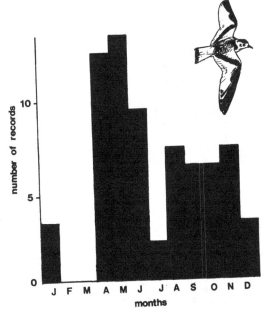

ber 1968 all the records were of single birds, many of which were immatures.

Present Status

An unusual annual visitor, usually on spring or autumn migration.
Apart from 1973 this species has been recorded in each year since 1970
with two or three records per annum. Seasonal distribution of all the
records is displayed in Figure 35 and shows spring migration with some
(immature) birds remaining well into June, and autumn passage with one
or two records in the early part of the winter. Numbers: remain small,
usually 1-5 but exceptionally 9 on the Ouse Washes in 1975 when nesting
was recorded (Carson, Cornford and Thomas 1977). Unfortunately, in
this case predation prevented hatching and the birds departed. The most

Little Gulls attempted to nest on the Ouse Washes in 1968.

favoured sites are the Washes (Ouse and Nene) together with, to a lesser extent, Ely beet factory.

CARSON C.A., CORNFORD G.A. and THOMAS G.J. Little Gulls nesting on the Ouse Washes. *Brit. Birds* 70. 1977.

Sabine's Gull   *Larus sabini*

An extremely rare vagrant.
Four records, listed below:

1 One record of a single bird at Quy around 1839. [Lack]

After the hurricane in October 1987 three records were received, the first for 150 years.

2 One on the Ouse Washes, 16 October 1987.
3 One on Midsummer Common, Cambridge, 20 October 1987.
4 One at Ely beet factory, 24 October 1987.

Black-Headed Gull   *Larus ridibundus*

Pre 1934

Evans stated that it was 'frequently observed as a rule towards winter'. Lack described it as a spring and autumn passage migrant, a winter visitor

Figure 36. Black-headed Gull – foreign recoveries of birds ringed in Cambridgeshire.

with a large roost on the Ouse Washes and in some years an oversummering species, particularly immatures. He noted the first breeding, since records began, at Burwell Fen in 1933.

1934-1969

At the beginning of this period it seems that the only known breeding colony was at Burwell Fen, and that wintering was on a smaller scale than at present. In 1938 there was a note of this species coming into Cambridge for the first time, in the hard weather. Breeding records in this period were centred on Ely beet factory (up to 200 pairs) where a colony was discovered in 1947, and in the early 1950s a colony at Peterborough sewage farm contained between 500 and 750 pairs. Spasmodic nesting in small numbers was noted at Cambridge sewage farm, Fulbourn Fen and on the Ouse Washes and attempted breeding was recorded at Wicken Fen in 1957. Roosting on the Ouse Washes, mentioned by Lack (see above), was regularly recorded with larger numbers occurring over the years up to a maximum of around 40 000 (January 1967).

Present Status

Present throughout the year as a winter visitor, passage migrant and breeding species.

Widespread, and in large parts of the county abundant out of the breeding season. Large flocks of 1000-3000 build up in favoured areas particularly around rubbish tips, while almost every open field has 1-50 foraging birds. The roost on the Ouse Washes varies in size but recent maxima have been of the order of 15 000-20 000. Howes (1976) showed that the main flight paths in the county are along the river systems: Cambridge-Ely, along the Old West to Huntingdon, and along the length of both the Ouse and Nene Washes.

Breeding Status

Only the colony at Ely beet factory is in regular use. The number of pairs is very variable but has decreased from 100-150 pairs in the early 1970s to around 20-40 at present. Nesting has been attempted on both the Ouse and Nene Washes in recent years but success has been limited.

Ringing Results

There have been several birds recovered in the county having been ringed abroad and vice versa and their countries of origin are displayed in Figure 36. These fit with the pattern described by MacKinnon and Coulson (1987) who estimated that up to 80% of birds of this species in southern

England in winter are Continental birds. These birds were all in Cambridgeshire in winter and on the Continent in summer with the exception of the Spanish bird which had moved further south a month after capture. An example of the speed of their movement is a bird ringed as a pullus in the USSR in early June 1983 which was controlled at Ely beet factory on 20 July of the same year. The work of the Gull Study Group shows some movement between feeding sites across the county over the winter.

HOWES M. The status of gulls in Cambridgeshire. *Camb. Bird Club Report* 50 1976.
MACKINNON G.E. and COULSON J.C. The temporal and geographical distribution of Continental Black-headed Gulls in the British Isles. *Bird Study* 34. 1987.

Common Gull   *Larus canus*

Pre 1934

Evans described this species as a winter visitor. Lack stated that it was a regular winter visitor with flocks of 60 or more in the fields. He reported that it was often associated with Black-headed Gulls but was far less common. He also reported that some immature birds were present in summer.

1934-1969

During the middle period this species remained a regular visitor out of the breeding season, both as a passage migrant and in winter. Numbers varied from one or two up to large flocks of several hundred. An extraordinary record of a bird taking fat from a bird-table in Soham can be explained simply by the date: January 1963!

Present Status

Recorded throughout the year, but mainly a passage and winter visitor with most of the birds in summer being immatures.
Roosts in winter on the Ouse Washes usually reach a maximum of between 1000 and 10 000, and flocks of 50-400 are reported from arable land; there are also some gatherings at rubbish tips but these are less numerous than those of the Black-headed Gull. This species is found all around the county.

Ringing Results

The Gull Study Group has had less success with this species than the Black-headed, but nevertheless evidence has emerged that the birds that winter in Cambridgeshire have similar origins. Birds have been caught from POLAND, WEST GERMANY, DENMARK and the USSR to date.

Lesser Black-Backed Gull *Larus fuscus*

Pre 1934

Evans reported that it 'occurs occasionally'. Lack stated that it was regular on both migrations; mid April to June flying north or north-east then returning mid July to October. He mentioned only one winter record (Gog Magog Hills, February 1910).

1934-1969

During this period the position was much as Lack stated. Records were restricted to the main passage periods, and although winter records became more regular with time, usually, even then, there were only two or three per annum. Even the reported habit of gulls feeding at rubbish tips failed to change this pattern. Numbers were generally small, usually 1-4, but some gatherings of up to 40, particularly in the north of the county, were noted and by the early 1960s one or two immatures were seen in summer.

Present Status

Recorded in all months of the year.

Its strongest status is as a passage migrant in both spring and autumn, although some immatures are seen in summer and some birds overwinter in appropriate areas around the Washes (Ouse and Nene). Numbers are not greatly different from above with 1-10 being seen together. There is, however, an increasing tendency for larger gatherings to occur with 40-50 on arable land particularly in August and September, and large flights (400 over Shepreth, 340 over Croxton in 1981 for example) to the Grafham Water roost. The main passage periods are end February to mid April and July to early November.

Herring Gull *Larus argentatus*

Pre 1934

Evans reported that it 'occurs occasionally'. Lack stated that it was a regular winter visitor in the north of the county, but very irregular in the southern parts.

1934-1969

Initially this species was a most unusual visitor with no more than one or two records each year. In the immediate post-war period it began to be recorded more often, although this may have been a result of more widespread observations. Largest numbers were on the Ouse Washes where a roost of up to 1500 was noted each winter. In the mid 1950s the habit of feeding at rubbish tips transformed its status and around both Milton (maximum 2100 in 1960) and Coldham's Lane, Cambridge (max 700 in 1959), large numbers gathered.

Present Status

Recorded in most months but predominantly in winter or on spring or autumn passage.

In the early 1970s the authorities began to cover the tips with topsoil thus restricting the amount of food available which resulted in a fall in numbers of most gull species to 100-200 on the tip sites and latterly smaller numbers still of this species. Seasonally Herring Gull can be seen in any month apart from the high-summer breeding period, but it remains most common in winter. Numbers: usually 1-20, but in favoured areas such as around the various washlands larger gatherings can occur and roosting is still regularly recorded on the Ouse Washes. Numbers in the extreme north of the county are still high where birds fly from inland feeding sites to roost on the coast and gatherings of 1000 or more occur around rubbish tips near Peterborough or, more recently, March.

Iceland Gull *Larus glaucoides*

A very rare vagrant.

Five records, all listed below:

1 An immature at Coldham's Lane rubbish tip, Cambridge, 24 January 1959. A similar bird was seen at Cambridge sewage farm on 14 February 1959 and was considered to be the same individual.
2 An adult flying low to the north at Cambridge sewage farm, 15 March 1961.
3 One flying south at Cambridge sewage farm, 9 February 1970.
4 One at Ely rubbish tip, 2 January 1975.
5 An immature on the Ouse Washes, 9 and 17 December 1979.

Glaucous Gull *Larus hyperboreus*

Pre 1934

One record of a bird shot at Caxton in November 1827. The body is preserved in the Cambridge University Zoology Museum. [Lack]

1934-1969

There were 15 further records in this period, about half (7) in Cambridge either at the sewage farm or at rubbish tips. Of the others four were on the Ouse Washes and four at other sites in the northern half of the county. All were of single birds, usually seen just once, and at least five were of immature birds.

Present Status

A rare winter visitor.

A further 12 records brings the overall total to 29. The habit of feeding at tips has possibly brought an increasing number of records recently since two-thirds of the total have been seen in the last 25 years. Milton tip, the Ouse Washes, Waterbeach GP and Stretham tip are the sites where birds have been seen more than once in this period. Numbers: usually single birds, two at Waterbeach in 1972 is the only exception. A large proportion are immature or sub-adult birds. Monthly distribution of records shows no particular preference but the peak, such as it is, occurs in February.

Earliest date: 14 November (1977 Milton)
Latest date: 24 April (1985 Ouse Washes)

Great Black-Backed Gull   *Larus marinus*

Pre 1934

Evans stated (as with most gulls) that it 'occurs occasionally'. Lack records it as a very rare winter visitor giving only three records: in 1824, 1826 and 1926.

1934-1969

As with the Lesser Black-backed Gull, scarcely had Lack published his work than records began to occur with some regularity, and in the 1940s this species was seen in almost every year with a maximum of 12 records per annum. In the 1950s this regular pattern was maintained though the species appears never to have been common. Nevertheless, the number of records increased with time and some large gatherings (30-40) were noted. In the 1960s feeding at rubbish tips led to many more records, higher numbers and roosting with up to 200 on the Ouse Washes.

Present Status

Recorded in every month of the year, but predominantly a winter visitor. Summer records are usually of immature birds.

Numbers: usually quite small (1-20), except at roosting sites where up to 500 (Ouse Washes 1975) have been counted. Otherwise gatherings only occur at rubbish tips. This species remains uncommon in the area south and west of Cambridge.

Ringing Results

A bird ringed at March tip in 1986 was found dead at Minsmere (Suffolk), in mid March 1987.

Kittiwake   *Rissa tridactyla*

Pre 1934

Evans declared that it 'usually occurs in hard winters' and added that it was never common, a position which is clear from the fact that Lack found only 12 records.

1934-1969

After six further spasmodic records up to 1953 it was seen annually. Numbers: usually 1-2 birds with records concentrated at or around Cambridge sewage farm or Ouse Washes during winter or passage periods. Records of large numbers such as the 200 counted over Milton and

Figure 37. Kittiwake – monthly distribution of all records.

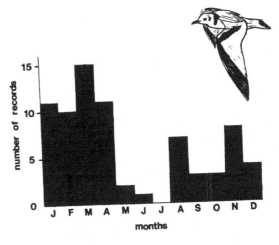

Waterbeach on 13 April 1969 were the result of birds being driven inland by coastal storms.

Present Status

A storm-driven visitor most commonly recorded in the period August to April.

This species was curiously unrecorded from 1971 to 1977 but subsequently has been recorded 2-5 times each year, very much as before except that more large groups have been noted. Of all the records a significant proportion involves either sick, oiled or dead or dying individuals and many are immatures. The seasonal distribution is shown in Figure 37.

Ivory Gull   *Pagophila eburnea*

An extremely rare vagrant.
One record only:

A single bird at Cambridge sewage farm, 27 February 1938.

Caspian Tern   *Sterna caspia*

An extremely rare vagrant.
One record only:

A single bird was seen on the Ouse Washes at Welches Dam, 5 July 1981; this was presumably the individual seen over Wisbech sewage farm on 3 July.

Sandwich Tern   *Sterna sandvicensis*

Pre 1934

Lack quoted a single record of two seen at Cambridge sewage farm in August 1931.

1934-1969

During this period there were 20 more records almost all during autumn passage. Numbers: usually 1-8 but exceptionally 20 were counted flying NNE over Wisbech in late August 1961. Most of the birds were seen or heard flying over, and the majority of reports came from Cambridge sewage farm.

Present Status

An unusual and irregular passage migrant.
There have been 26 additional records since 1970, most of which are similar to those analysed above. The seasonal distribution of all the records is shown in Figure 38. Numbers: mostly 1-20, exceptionally 70 over Wisbech in September 1970 and 50-100 over Milton in September 1983. Almost all records are of birds seen or heard flying over except for two which roosted with gulls at Histon on 9 September 1976 after heavy gales.

Ringing Results

A single bird picked up dead on the Ouse Washes on 28 April 1981 had been ringed as a juvenile at Strangford Lough, Northern Ireland, in July 1970.
Earliest date: 28 March (1987 Littleport)
Latest date: 16 October (1987 Ouse Washes)

Roseate Tern   *Sterna dougallii*

An extremely rare vagrant.
One record only:
A single bird was seen on the Ouse Washes (Purl's Bridge), 11 May 1981.

Common Tern   *Sterna hirundo*

Pre 1934
Evans described this species as a straggler, but by Lack's time it occurred

Figure 38. Sandwich Tern – monthly distribution of all records.

fairly regularly on both migrations with parties of up to 50 over Chesterton Ballast pits.

1934-1969

Recorded annually initially at/over Cambridge sewage farm on spring or autumn passage. Numbers: usually 1-5 but not uncommonly 20-30. Strength and duration of passage varied from year to year and there was no seasonal preference. In 1956 birds summered at Peterborough sewage farm, and in 1963 at Ely beet factory, without breeding. However, by the late 1960s nesting was recorded on the larger gravel pits in nearby Huntingdonshire. In this period the number of records in Cambridgeshire began to increase.

Present Status

Summer resident and passage migrant.

Numbers: generally on passage around 1-10 but with the increased volume of records so the strength and duration of passage has increased. The main passage periods are mid April to end of May, and throughout August and September. There are often identification problems with Common and Arctic Terns particularly when birds are seen only fleetingly or in such large numbers that each individual cannot be identified; therefore many of these records refer to 'commic' terns.

Breeding Status

The first record was at Chatteris GP where a pair nested in 1971 but were flooded out. In 1975 two pairs bred at Fen Drayton GP, a site which has been used regularly since with a maximum of four pairs. Breeding (2-4 pairs) has also been recorded at Eldernell (1982) and Whittlesey West pits (1981), and on the Ouse Washes (1982 and 1984).
Earliest date: 28 March (1981)
Latest date: 6 December (1953 Landbeach GP)

Arctic Tern   *Sterna paradisaea*

Pre 1934

Evans quoted a single record from Jenyns (1834) and suggested underrecording, with which Lack was in agreement. Lack also reported four further records of single birds seen at Cambridge sewage farm on 6 May, 26 May and 11 August (when there may have been two individuals) all in 1930 and on 12 July 1932.

1934-1969

One or two records in most years although due to the difficulty of separation between Common and Arctic many are assumed to be Common; thus it is possible that Arctic has always been under-recorded. Numbers: always less than seven and usually one or two. The only site mentioned regularly was Cambridge sewage farm. The only unusual record was of seven flying over Witcham on 28 June 1966 following a north-west gale.

Present Status

A regular passage migrant in spring and autumn.
Recorded slightly more often with a recent seasonal bias to spring records (mid April-mid May) than autumn (end July to end September). Numbers: remain much lower than for Common Tern, usually 1-10, only one record of 20, on the Ouse Washes in early May 1981. The Ouse and Nene Washes, Ely beet factory and Fen Drayton GP are the most favoured sites.
Earliest date: 5 April (1974 Ouse Washes)
Latest date: 29 October (1949 Ouse Washes)

Little Tern   *Sterna albifrons*

Pre 1934

Lack quotes two records: one shot at March in May 1850 and one at Cambridge sewage farm in May 1930. Two other records at Cambridge sewage farm were on 2-6 August 1930 and late May 1931.

1934-1969

There were 27 records in this period of which just over half were at, or near, Cambridge sewage farm/Milton GP. Of the remainder: four were on the Ouse Washes, four in the Littleport-Ely area and three at Peterborough sewage farm. Numbers: usually single birds but exceptionally seven at Littleport, 13 June 1956.

Present Status

An unusual passage migrant.
There have been a further 14 records since 1969 with much the same criteria as mentioned above. Six records were on the Ouse Washes, five in the Littleport-Ely area, one at Fen Drayton GP and one at Waterbeach GP. Numbers: usually singles but exceptionally six in April/May 1970 and five

in May 1978 (both on the Ouse Washes). Seasonal distribution is shown in Figure 39 and gives a clear indication that over half of all the records occur in May. Records show no sign of either increase or decrease.
Earliest date: 19 April (1952 Cambridge sewage farm)
Latest date: 27 September (1957 Milton)

Black Tern   *Chlidonias niger*

Pre 1934

Jenyns noted flocks near Bottisham in 1824 and stated that it was very abundant, he also quoted a record of nesting near Bottisham in 1824. Evans also mentioned flocks at Gamlingay in 1831. Lack stated that it was a regular spring and autumn passage migrant over any open water from late April to early June and late July to September, 1-10 being seen together.

1934-1969

Recorded with great regularity in spring and autumn particularly at Cambridge sewage farm where large numbers (up to 70) gathered in some years especially in spring (see Fig.40), at Peterborough sewage farm (up to 65), and on the Ouse Washes. The timing, strength and duration of passage varied from year to year but generally more individuals passed through in spring than autumn. In some years, however, observed migration was poor.

Figure 39. Little Tern – monthly distribution of all records.

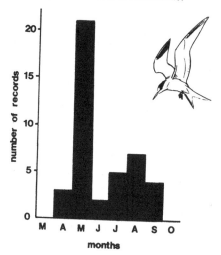

Present Status

A regular spring and autumn passage migrant, with some over summering.

Seen mainly on the Ouse Washes and at Ely beet factory but can occur at any of the wetland sites: Nene Washes, Fen Drayton, Landbeach, Milton and Waterbeach GPs. Numbers: can be up to a maximum of 139 (on the Ouse Washes in May 1981) but generally 1-20 in any one place at one time is usual. The main passage periods are mid April to end of May, and end of July to beginning of October, but these periods remain variable and often migration takes place in waves.

Breeding Status

Breeding was first reported in 1966 on the Ouse Washes (Cottier and Lea 1969) and after several nests had been constructed two pairs remained, both being successful. In 1969 on the Ouse Washes up to six pairs were present one of which was successful; however, not all these nests were in Cambridgeshire. In 1975 out of two pairs one was successful, but in other

Figure 40. Black Tern – extreme spring migration at Cambridge sewage farm in May 1948.

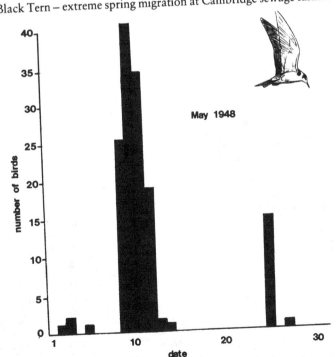

years, despite the presence quite often of oversummering birds, breeding has not been attempted.

Earliest date: 9 April (1952, 1956 Cambridge sewage farm)

Latest date: 21 October (1983 Ouse Washes)

COTTIER E.J. and LEA David, Black-tailed Godwits, Ruffs and Black Terns breeding on the Ouse Washes. *Brit. Birds* 62. 1969.

White-Winged Black Tern   *Chlidonias leucopterus*

A very rare vagrant.

[Two records from Livesey (Field 1911) were rejected by Lack on the grounds of identification.]

Five records, listed below:

1 One along the River Nene, 8 September 1963.
2 Two or three birds irregularly present on the Ouse Washes, 17 August-23 September 1968.
3 An adult at Ely beet factory, 12-23 August 1970.
4 An adult in summer plumage at Upware, 29 May 1981.
5 One seen flying along the Nene Washes, 7 June 1985.

Black Terns nested on the Ouse Washes in 1966.

Guillemot   *Uria aalge*

A rare vagrant.

Five plus records, all listed below:

1. One from Burwell sent to Farren, September 1904. [Lack]
2. One from Meldreth sent to Farren, November 1910. [Lack]
3. One unable to fly near Abington, 26 February 1935.
4. Two seen from a bedroom window flying over King's College 'Backs' at 7.00 a.m. on 9 March 1959. The observer was said to know the species well!
5. In 1983 there was a massive auk 'wreck', particularly in the north of the county along the River Nene. In a stretch from Whittlesey to Wisbech sewage farm there were five or more live and 14 dead birds found from 11 February to the end of the month and on 14 March 25 bodies were counted (including some of those mentioned previously). Elsewhere, on the Ouse Washes live birds were seen from 10 to 14 February and two corpses were discovered. The last live bird was at Manea on 1 March.

(As it is impossible to number these records all subsequent will not be numbered.)

One on the River Nene, 9 February 1986.

Brunnich's Guillemot   *Uria lomvia*

An extremely rare vagrant.

One record only:

A single bird was obtained at Guyhirn in January 1895. [Lack]

Razorbill   *Alca torda*

An extremely rare vagrant.

Three records, listed below:

1. One at New Wimpole in October 1835 is preserved in the Cambridge University Zoology Museum. [Lack]
2. One at Wisbech on 12 February 1983.
3. Two at Guyhirn on 13 February 1983.

Records 2 and 3 were as a result of the auk 'wreck' in February 1983 (see Guillemot).

Little Auk   *Alle alle*

Pre 1934

Evans stated that it was to be 'met with in severe winters'. Lack clarified this by stating that it occurred in winter after severe gales. Up to 1934 there were about 30 records and of these 17 were since 1900. Most records were of birds found dead.

1934-Present Status

A rare vagrant.

All records in this period are listed below:

One at Littlington, 11 February, and one at Shippea Hill, 12 February 1950 were both part of a national 'wreck' following south-westerly and westerly gales described by Sergeant (1952).

One found dying in the Botanic Garden in Cambridge, 1 February 1953.

One found exhausted at Guyhirn, 23 October 1955.

In November 1957 one at Cambridge sewage farm, on the 10th, one at Six Mile Bottom on the 11th and three records in the Littleport area late in the month.

One found dead at Soham in the second week of November 1958.

In 1974 a 'wreck' on the east coast led to one at Welney, 24 October, and one at Wicken Fen along the lode, 1 November.

In February 1983, one at Wisbech, on the 8th, one at Waterbeach GP, on the 13th, one found dead at Harlton, on the 21st, and another at Guyhirn.

One found on the Devil's Dyke at Burwell, 7 November 1984.

SERGEANT D.E. Little Auks in Britain 1948-1951. *Brit. Birds* 45. 1952.

Puffin   *Fratercula arctica*

A rare vagrant.

Twenty records to date, listed below:

1 One near Ely, February 1852. [Lack]
2 One near March, November 1893. [Lack]
3 One from Over sent to Farren, February 1903. [Lack]
4 One from Soham sent to Farren, February 1907. [Lack]
5 One from Abington Piggots sent to Farren, January 1918. [Lack]
6 One found dead near Royston, 23 November 1937.
7 One found dead at Wilburton, 31 January 1953.

8 One found dying in the Botanic Garden, Cambridge, 12 February 1953.

9 Two on the River Ouse at Over, 28 November 1953, of which one was shot the following day and the other died in December.

10 One found dead at Wisbech St Mary, 23 October 1955.

11 One shot at Oxlode on the Ouse Washes, 2 November 1958.

12 One dead at Over, 23 November 1958.

13 One on the Nene Washes, 26 November 1958.

14 One seen oiled on Quy Water, 7 December 1958.

15 An immature on the River Nene at Whittlesey, 12 November 1961.

16 An immature picked up alive at Barrington, 21 October 1962.

17 Two at Over staunch from 19 to 21 October 1969.

18 An immature, found dead on the Cambridgeshire side of the Great Ouse, 2 February 1974.

19 An immature found alive at Coldham's Lane, Cambridge, 17 February 1979 was cared for and later released at the Nene mouth.

20 One at March on 10 February 1983.

Palla's Sand Grouse   *Syrrhaptes paradoxus*

An extremely rare vagrant.
Numbers of this species were seen in the two irruption years of 1863 and 1888. [Lack]

LANDBIRDS

As in most counties the level of interest in this section of birds has been more limited. The reasons for this are many but most obvious are the attraction of the species that have already been discussed and the relative

This immature Puffin appeared at Over Staunch in October, 1969.

abundance of landbirds leading to them being overlooked (ignored!) by birdwatchers. It is also unfortunate that the Common Bird Censuses that have been carried out in the county have not been continuous, and in addition the census carried out by undergraduates at Barton Farm, Coton (Watmough 1973, Bucknell 1977) was not entirely typical of the county since the site underwent such considerable change, partly agricultural, partly housing development, but mainly associated with the building of the western by-pass. A possible measure of populations is available from the ringing returns of the Wicken Fen Group which has worked continuously since 1968 but with variable effort and success. Since the aim was to catch acrocephalus warblers most other birds were caught coincidentally therefore the changes in the numbers could reflect the pattern of population variations, and these are presented for certain species. Indicators of population densities of common birds for two habitats, suburban (Harper) and woodland (Conder), are included where suitable. Easy (pers. comm.) has pointed out that there is a considerable difference in the distribution and status of common species between the upland and the fenland areas. There is very little published information on this aspect. He states by way of example that in fenland you will see many more Stock Dove than Blue Tit, making it difficult to give precise statements on such species.

The following references are cited throughout the section on Landbirds:

BUCKNELL N. The Common Bird Census at Coton. *Camb. Bird Club Report* 51. 1977.
CONDER Peter. Birds. In Rackham O. *Hayley Wood, Its History and Ecology.* CAMBIENT. 1975.
HARPER D.C.G. Birds of the University Botanic Garden Cambridge. *Cambridge. Univ. Bot. Gdn.* 1982.
WATMOUGH B. *The Common Bird Census at Barton Farm, Coton, 1967-1973. Camb. Bird Club Report* 47. 1973.

Stock Dove   *Columba oenas*

Pre 1934

Evans described this species as 'abundant', while Lack stated that it was well distributed throughout the county though not as common as the Wood Pigeon. He considered it commoner in winter when large flocks can occur.

1934-Present Status

A thinly distributed resident found across the county.
Often associated with Wood Pigeon in feeding flocks, but records show

that some Stock Dove gather separately with counts from 20-100 in areas all around the county. O'Connor and Mead (1984) suggest a national decline from the levels of the 1950s due to the combined effects of pesticide poisoning and Dutch Elm disease. Although in Cambridgeshire it has not been very thoroughly surveyed since the decline of the elm, its main nest site, in the 1960s, it has favoured other sites, e.g. deserted cottages and other buildings in fenland, and holes in chalk cliffs as at Cherry Hinton. In the southern part of the county it is rather poorly recorded being present throughout if rather thinly distributed.

Breeding Status
The *Breeding Atlas* (Sharrock) shows this species to be widespread with the exception of a section on the eastern boundary.

Ringing Results
A bird ringed at Soham in June 1961 was found at Tilbury, (Essex), in January 1962.

O'CONNOR R.J. and MEAD C.J. The Stock Dove in Britain 1930-1980. *Brit. Birds* 77. 1984.

Wood Pigeon   *Columba palumbus*

Pre 1934
Evans stated that it was common. Lack described it as extremely abundant and added that in winter birds were usually found in large flocks and roosts as at Madingley and on the Gog Magog Hills. Breeding was abundant.

1934-1969
Common and widespread. The Ministry of Agriculture destroyed 1743 eggs and 1419 young in a four square mile area around Bottisham between 19 July and 20 August in 1954. This would suggest a density of five pairs per acre. The 1962/63 winter was said to have had a drastic effect on breeding numbers the following summer but the recovery seems to have been very rapid! Murton (1965) conducted most of the research that went into his monograph in Cambridgeshire and therefore his work is directly applicable.

Present Status
A most widespread and abundant resident.

Many records of flocks of 1000 or more are received. With the increase in arable land that has taken place this century this species has assumed pest proportions and almost every field is visited at one time or another. Roosts occur almost anywhere (I recently counted about 1000 birds at Cherry Hinton Cement Pit!) and all areas of woodland are used, often by numbers which seem inordinately large for the number of trees. Despite regular shooting of feeding and roosting flocks there seems to be little drop in numbers. In Hayley Wood in both summer and winter Conder (1975) found this species to be the most common relative to all others.

Breeding Status

A widespread and extremely abundant breeding species. The *Breeding Atlas* (Sharrock) suggests a density of 50 pairs per square kilometre in lowland arable areas. In the Botanic Garden there is a density of about 15 pairs (Harper 1982).

Ringing Results

Recoveries suggest that most of the Cambridgeshire population are local (Murton) but two birds have travelled as far as FRANCE these were ringed as fledgelings and recovered in their first year of life. However, most other recoveries are either very close to the point of ringing or at least within the south or east area of England.

MURTON R.K. *The Wood Pigeon*. Collins 1965.

Collared Dove   *Streptopelia decaocto*

1934-1969

Hudson (1965) analysed the spread of this species through Britain and Ireland, noting the first record in Lincolnshire in 1952 and the first breeding record in Norfolk in 1955. Since these two counties are both adjacent to Cambridgeshire it is surprising that the first Cambridgeshire record was not until 1960. However, Easy (pers. comm.) states that records in the 1950s were considered to be misidentifications of feral species!

1960 First record, when two birds were seen at Adams Road Bird Sanctuary, Cambridge, from 20 November to 4 December.

1961 There was a breeding record (Vine 1962) at Littleport where a pair raised three broods.

1962 A second breeding site in Shelford was used.

1963 Eight pairs were breeding.

1964 Breeding was to be found in nine widespread villages but none in Cambridge or south of the city.

1965 Completed the colonisation.

1967 A widespread breeding species (300+ pairs).

1968 Three years after first breeding in the city there was a population of 25 pairs in Cambridge.

Present Status

A widespread and abundant resident.

Having colonised the county as a whole, numbers continued to increase until they reached saturation point around 1972 or 1973, just over ten years after the first record. Throughout the last 15 years records of flocks and roosts up to 150 have not been unusual and smaller gatherings of 20-40 are very common.

Breeding Status

By the time that the *Breeding Atlas* (Sharrock) was completed it was a widespread and abundant breeding species, a good example being that in 1974 70 pairs were counted on the Ouse Washes. As with the Wood Pigeon the season is protracted and several broods are raised. In 1981 seven pairs bred in the Botanic Garden (Harper 1982).

Ringing Results

A bird ringed in HOLLAND in January 1966 was found dead at March in March 1967. The only other recoveries are of three birds ringed during the winter of 1971 at a gathering in a grain store in Manea which were found in Perthshire in January 1974, Ayrshire in January 1974 and in Lancashire in February of the same year.

HUDSON R. The spread of the Collared Dove in Britain and Ireland. *Brit. Birds* 58. 1965.
VINE A.E. First county nesting of Collared Doves. *Nature in Cambridgeshire* 1962.

Turtle Dove *Streptopelia turtur*

Pre 1934

Evans described this species as a common summer visitor, often very plentiful. Lack added that it was widely distributed and arrived around the end of April.

1934-Present Status

Widespread and common summer visitor.

There was little change in status and it remained a common summer resident present from around the middle of April until the middle of October right across the county. Some moderate flocks (up to 150) build up after breeding and before migration, or on feeding grounds. Since 1980 some decline has been evident, more especially in the breeding population.

Breeding Status

It is shown in the *Breeding Atlas* (Sharrock) to be a widespread breeding species. For example, 60 pairs were counted at Ditton Park Wood in 1972, and on the Ouse Washes there were 85 pairs in 1974 and 33 in 1975.

Ringing Results

Three birds, one ringed in Shelford in July 1937, one at Madingley in June 1951 and the third at Wicken Fen in May 1975, were recovered in FRANCE in May 1939, May 1954 and September 1976 respectively. A recovery in MALI in April 1976 of a bird ringed at Wicken Fen in July 1973 was confirmation of wintering/migration area. A bird ringed at Wicken Fen in June 1984 was shot at Badajoz, SPAIN, in September 1985, presumably on departure migration.
Earliest date: 4 April (1985)
Latest date: 16 November (1957 Cambridge)

Ring-Necked Parakeet   *Psittacula krameri*

Some individuals were reported prior to 1976 but not published since at the time this species was not considered a likely addition to the British List. A feral species at present only rarely recorded.
Nine records since 1976, all listed below:

1 One at Horningsea on 19 October 1976.
2 One at Ely beet factory, 4 November 1976.
3 One at Wicken Fen, 21 May 1977.
4 One flying east at Dimmock's Cote on 30 April 1978.
5 One at the Botanic Garden, Cambridge, on 13 July 1979.
6 A male flying north-east over the Ouse Washes, 18 August 1979.
7 One seen several times around Cambridge, 1984.
8 A female on Waterbeach Fen, 13 October 1985, was being chased by a Black-headed Gull.
9 One at Kneesworth, near Bassingbourn, 2 November 1986.

Since this species is a recent addition to the British List further records can be expected and it has not therefore been given rarity status in this list.

Cuckoo   *Cuculus canorus*

Pre 1934

Evans stated that this species was 'very common in summer'. Lack described it as a summer resident which arrived in the third week of April and was to be found throughout the county, particularly in the fens.

1934-Present Status

A common and widespread summer resident.
There is no firm evidence of change in the status of this species over the last 50 years despite considerable changes to the environment. Found in most villages, it has not really spread into suburbia but at its favoured haunts it remains plentiful. Much of the parasitism is based on Reed Warblers at such sites as the Ouse Washes, Fulbourn and Wicken Fens and Fowlmere watercress beds. On the Ouse Washes counts made in June 1978, 1980 and 1985 noted 17, 20 and 22 birds respectively.

Breeding Status

According to the *Breeding Atlas* (Sharrock) it is well distributed with quite large numbers at suitable sites. In 1985, apart from 22 on the Ouse Washes (see above), there were 12 females at Wicken Fen, 12 adults on the Nene Washes, 4+ at Fen Drayton GP and two females at Fowlmere.

Ringing Results

A bird ringed as a pullus at Wicken Fen in July 1985 was reported 'sick now released' at Arnhem in the NETHERLANDS one month later having moved due east.
Earliest date: 30 March (1983)
Latest date: 13 October (1969 Milton)

OWLS

Like other birds of prey, owls suffered badly during the organo-chlorine disaster of the 1960s, the Barn Owl in particular. This species has been slow to recover with the additional stress placed upon agricultural habitat in recent years. The Tawny Owl, however, seems to be flourishing with the growth of human habitation and suburbia. The Little and 'Eared'

Owls remain in their traditionally favoured haunts but probably in reduced numbers.

Barn Owl   *Tyto alba*

Pre 1934

Evans described this species as abundant and Lack gave a full account of its status saying that it was found all over the county. In 1932 a status survey throughout England and Wales showed that there were 96 pairs in Cambridgeshire and it was fairly numerous across the county but particularly south of a line from Newmarket to St Ives; however, all observers reported it to be decreasing with some previously occupied nest sites empty. Farren told Lack that it was the most favoured species for collectors.

1934-1969

A record from 1944 shows how birdwatching in Cambridgeshire has changed; in February one of a pair of white-breasted birds at Cambridge sewage farm was killed by one of a pair of Peregrine Falcons! There were many records through this period until the early 1960s when the effects of toxic chemicals reduced the population although it remained in its stronghold in the fenland. Osborne (1982) showed that Dutch Elm disease had an extremely detrimental effect on populations of many birds simply because their nest sites were lost and he lists this species among them. In 1964 there were records from only 15 parishes; eight in north Cambridgeshire, six in central Cambridgeshire and one in the south. By 1968, however, there were signs of a slow recovery with the number of parishes in which Barn Owls were recorded up to 25.

Present Status

A locally distributed resident.
Widespread in the area north of Cambridge with records from many parishes, particularly in the fenland. Unfortunately a significant proportion are found dead as a result of traffic accidents. South of Cambridge this species is a relatively uncommon sight but recent reports suggest that a few birds are to be found in the more open arable areas.

Breeding Status

After a period of scarcity in the early 1970s, shown quite clearly in the distribution in the *Breeding Atlas* (Sharrock) with only six breeding records in the county in 1971, a recovery has led to a more healthy population (c.15 sites in 1975). However, this remains considerably fewer than before the traumas of the 1960s.

OSBORNE P. Some effects of Dutch Elm disease on nesting farmland birds. *Bird Study* 29. 1982.

Little Owl   *Athene noctua*

Pre 1934

One was reported in 1867 prior to the introduction of the colony by Lord Lilford. Evans stated that it was a rare straggler, probably from the colony established by Lord Lilford in Northamptonshire. Lack considered it to be a widely distributed and not uncommon resident. Farren received his first specimen in 1898, and this species then increased in Cambridgeshire at a great rate and was well established by 1905.

1934-1969

The status remained much as stated by Lack, in 1946 for example, 23 were destroyed on a Thriplow estate in traps set for stoats. Like the Barn Owl this species is more a bird of open country and is therefore similar in distribution.

Present Status

A thinly distributed resident.
In recent years there have been records from around 30 to 40 parishes, and it is more common in the south of the county than the Barn Owl. It shows little sign of moving into more suburban habitat, remaining very much a bird of open countryside. Some daytime roost gatherings have been recorded with 3-6 birds involved.

Breeding Status

The *Breeding Atlas* (Sharrock) shows that it breeds throughout the county but with a rather patchy distribution. In 1975 pairs were reported from 26 parishes.

Ringing Results

Three birds ringed as nestlings, two in Milton and one in Coton, were all recovered within a few parishes of their ringing location, within six months of their ringing.

Tawny Owl   *Strix aluco*

Pre 1934

Evans described this species as 'decidedly uncommon'. By Lack's time it was not numerous, although found throughout the county. Common in Cambridge, it was said to be increasing.

1934-1969

Many records were received each year but giving the clear impression that where the Barn Owl was common the Tawny was scarce and where the Tawny was present the Barn Owl was not. Cambridge and its immediate suburban area provided many reports.

Present Status

A generally well-distributed resident not uncommon in woodland and suburban areas but less common to the north of Cambridge. Records are regularly received from suburban areas where this species is moderately conspicuous by its calling during the hours of darkness. It is found in between 15 and 20 parishes but often several pairs per parish.

Breeding Status

The *Breeding Atlas* (Sharrock) shows it to be absent from much of the northern (fen) part of the county but widespread elsewhere. In the areas where it predominates many pairs can breed close together; for example, since 1980 the annual owl count at Wicken Fen has revealed a population of 10-13 pairs on, or close to, the reserve.

Ringing Results

A bird, ringed at Fowlmere in April 1973, was recovered at Shepreth in 1978, which tends to show the sedentary nature of this species.

Long-Eared Owl   *Asio otus*

Pre 1934

Jenyns considered this species to be rare. Evans described it as 'local' but present in the woods between Babraham and the Gog Magog Hills. Lack summarised its status as a sparse resident. A pair or two nested in Madingley with a few other pairs recorded in the fens or in woods at the periphery of the county

1934-1969
It was recorded regularly throughout this period particularly at certain sites: Wicken Fen (up to two pairs), Eldernell, Hildersham, the Ouse Washes, Balsham, Bottisham, Babraham, Devil's Dyke, etc.

Present Status
Resident with a scattered distribution. Probably a winter visitor and passage migrant in addition.

In some years (1975) there are minor invasions in autumn leading to large numbers, for example ten at Coveney in January 1976 and eight at Quy Fen in March 1976. Gatherings and roosts are now reported more regularly with up to ten individuals at certain sites.

Breeding Status
The *Breeding Atlas* (Sharrock) shows a patchy distribution nationally of which Cambridgeshire is a good example. Where this species is found there often exists a small pocket of birds as is the case at Wicken Fen where the annual owl count shows a breeding population of between three

and six pairs. Nesting occurs regularly at other places such as Swavesey, and the Ouse and Nene Washes.

Ringing Results

One ringed in June 1972 at Wicken Fen was found in Bottisham in April 1977 and one ringed at Swavesey in May 1974 was found at Twyford (Hants) in August 1975.

Short-Eared Owl   *Asio flammeus*

Pre 1934

Jenyns noted that it bred once or twice, and Evans added that it was regular on migration in autumn. Lack stated that 2-5 pairs bred regularly at Wicken Fen and that in winter it was a regular visitor, more numerous in some years than others, with parties of up to 12 in 1930-31.

1934-1969

In the area north of Cambridge and particularly in the fens this species was regularly recorded in winter. Large gatherings occurred at certain sites and up to ten were quite commonly seen, even up to 20 in some exceptional years. Favoured sites included the Ouse and Nene Washes and Fulbourn and Wicken Fens. Breeding took place in many years and the sites used included Wicken Fen (up to 2 pairs), Fulbourn Fen (1-2 pairs) and the Nene Washes (1-2 pairs).

Present Status

Predominantly a winter visitor, but still an irregular breeding species. Recorded on both the Ouse and Nene Washes where many indiviuals are scattered along the whole length; and at certain sites where roosts of birds that hunt locally over arable land form in some years. Numbers vary considerably from year to year. For example in 1970 (a good year) there were 24-36 on the Ouse Washes, up to 21 on Fulbourn Fen, 10 on the Nene Washes and singles at many other sites, whereas in 1975 (an average year) although there were records from 28 sites the maxima were five on the Ouse Washes, 12 on Fulbourn Fen and eight on the Nene Washes; these being more normal numbers.

Breeding Status

An irregular breeding species. Since 1970 records have been received from the Ouse (?1971, 4 pairs in 1973) and Nene (2 pairs in 1984) Washes. In other years breeding has been suspected but not proven.

Nightjar   *Caprimulgus europaeus*

Pre 1934

Evans recorded this species as a summer migrant, local and nowhere common. Lack went a stage further and described it as very local. However, he listed nesting sites on the Gog Magog Hills, Shelford, Cottenham, Gamlingay and near Chippenham. Farren reported that he received specimens in most years.

1934-1969

Although there were annual records until 1951 there was a dramatic decline throughout this period. Breeding information (often retrospective) suggested that nesting continued at Chippenham until 1958 and at Wood Ditton 2-3 pairs were said in 1955 to breed annually. A record of a pair at Shepreth in 1952 is accompanied by a note to the effect that it was known regularly at Haslingfield and Barrington up until 1949. After 1958 there were just two further records in this period, one on the Ouse Washes on 18 August 1968 and one at Cherry Hinton, 17 May 1969.

Present Status

An extremely uncommon passage migrant.
The only record since 1970 is a bird seen flying at dusk at Wicken Fen, 12 May 1980.
Earliest date: 14 May (1944 Madingley)
Latest date: 23 September (1944 Cambridge sewage farm)

Swift   *Apus apus*

Pre 1934

Evans described it as a plentiful summer visitor, and Lack reiterated this view noting that huge flocks occurred at Cambridge sewage farm in the evenings.

1934-Present Status

A widely distributed and very common summer resident.
Present from the last days of April to the end of August, with some stragglers into October. Gatherings of several hundred are reported in most years at one or two sites on still days or evenings, such as 600 seen feeding over a pea field south of Cambridge on 22 July 1975 and exceptionally 2500 over the Ouse Washes on 21 July 1974. In a fascinating

paper Darlington (1951) showed that, using 57 cyclists as observers in 19 units, it was possible to display clear movements into the county in early May. He also demonstrated that while in clear conditions the movement was on a broad front, in periods of wet weather birds followed the lines of the various rivers.

Breeding Status

Commonly found breeding in all villages and towns but due to a shortage of suitable sites rarely in open countryside.

Ringing Results

One ringed at Milton in May 1961 was found at Binbrook (Lincs) in July at a time that suggested that it passed through Cambridgeshire on its way to its breeding site. A single bird ringed at Downham Market (Norfolk) in June 1966 and recovered at Wicken Fen in May 1968 was almost certainly doing the same.

Earliest date: 15 April (1951)
Latest date: 27 October (1960 Cambridge)

DARLINGTON Arnold. The use of mobile observers in the study of patterns of migration. Brit. Birds 44. 1951.

Alpine Swift   *Apus melba*

An extremely rare vagrant.
Three records, all listed below:

1 One mentioned by Jenyns killed between Cambridge and Grantchester in September 1838 is to be found in the Cambridge University Zoology Museum.
2 One seen in Cambridge in May 1844.
3 One at Cherry Hinton, 11 September 1969.

Kingfisher   *Alcedo atthis*

Pre 1934

Evans stated that this species was 'not uncommon' and Lack described it as a sparsely distributed resident rather rare in the fens in summer.

1934-1969

Regularly reported throughout the county with breeding records from many sites (e.g. in 1956 12 pairs were said to be nesting in south Cambridgeshire alone) until 1963. In that year it was virtually exterminated by the effects of the awful winter there were only two records, one in July and another in September. By 1964 there was a sign of a return with a possible nesting record at Chippenham and four/five autumn records. There was no breeding reported in 1965 but by 1966 possible nesting was recorded in five parishes. It was not until 1969 that records returned to the level of those pre 1963.

Present Status
Widely reported resident from many areas of waterway and gravel pits. Numbers are probably highest in autumn but whether this is a result of immigration or breeding success is not known.

Breeding Status
This species breeds regularly at traditional sites around the county. The *Breeding Atlas* (Sharrock) shows a slightly patchy distribution in the northern part of the county.

Ringing Results
Three birds ringed at Fowlmere were recovered respectively on the north Hertfordshire border, Bury St Edmunds and Mepal, the first and last being within six months of ringing, the middle bird over three years later. Birds ringed as juveniles at Wicken Fen were found at Hauxton, within three months of ringing, at Witham (Essex), four months after ringing, at Cambridge (same delay), and at Burwell three months later. This species is obviously very vulnerable during the post-juvenile period. The most interesting recovery is of a bird ringed at Renishaw (Derbys) in July 1979 which was found at Wisbech the following September.

Roller *Coracias garrulus*

An extremely rare vagrant.
Two records, both 19th century, listed below:
1 One shot near Oakington in October 1835. [Lack]
2 One at Burwell Fen, June 1884.

Hoopoe *Upupa epops*
Pre 1934
Lack quoted eight records as follows:

One from Newmarket in June 1826.

An undated record from Royston

One from Littlington in September 1841. (These were all mentioned by Jenyns.)

One near Great Abington in 1856,

One near Ely, late autumn 1868.

Farren received specimens from Linton in November 1898, and from Pampisford about 1927.

Evans reported one near Cambridge about 1912.

1934-1969

A further nine records were noted in this period. All were of single birds with the exception of 1-2 in Cambridge from the end of April to early May and two (the same two?) at Over, 2 May 1963. These records came from around the county, mainly in spring.

Present Status

An uncommon vagrant.

Since 1970 there have been 15-16 records, all of single birds. This gives an overall total of just over 30 records. In the case of so distinct a bird, when two records fall close together the possibility of the same individual being involved cannot be ruled out. The recent increase in records is most likely a reflection of the increased observation rather than any population effect. Almost all the records have been in spring (April/May) as can be seen in Figure 41 which shows the monthly distribution of all the records since

Figure 41. Hoopoe – monthly distribution of all records.

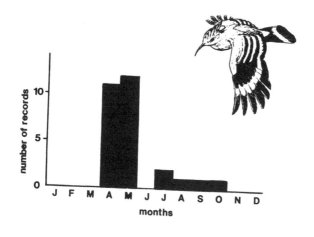

1934, and this agrees with the national status as summarised by Sharrock (1969).
Earliest date: 14 April (1964 Cambridge, 1974 Swavesey)
Latest date: November (1898 Linton)

SHARROCK J.T.R. Scarce migrants in Britain and Ireland 1958-1967. *Brit. Birds* 62. 1969.

Wryneck   *Jynx torquilla*

Pre 1934
Jenyns considered this species common. Evans described it as a 'somewhat rare summer visitor'. Lack stated that it was an occasional passage migrant. Farren told Lack that it was common around Cambridge up until the 1880s and that it ceased to breed at the turn of the century. Records from the Cambridge 'Backs' in 1909 and 1910 suggested breeding.

1934-1969
The national decline of this species, outlined by Monk (1963) and Peal (1968), seems to have begun in the early 1900s and Cambridgeshire was one of the first counties to be deserted. Therefore it is hardly surprising that during this period there were only six records, all of single birds with the exception of a pair on the 'Backs' in April 1968. Four of the records were in spring and the other two in autumn. The reason for the decline of this species remains unclear and may be due to a combination of factors.

Present Status
An unusual and irregular passage migrant.
There have been 29 further records of which, remarkably, only two were spring birds all the others being in August or September. The only record of two birds was at Whittlesey in September 1970, otherwise all birds were singles. There is no particular locational bias except towards Cambridge which is almost certainly merely an observer bias.
Earliest date: 21 April (1968 Cambridge)
Latest date: 4 October (1976 Sawston)

MONK J.F. The past and present status of the Wryneck in the British Isles. *Bird Study* 10. 1963.
PEAL R.E.F. The distribution of the Wryneck in the British Isles 1964-1966. *Bird Study* 15. 1968.

Green Woodpecker   *Picus viridis*

Pre 1934

Evans described this species as 'not uncommon and in some districts abundant'. Lack stated that it was a widely distributed resident including in the fens but was nowhere numerous. Farren, confirmed by Evans, gave Lack the impression that it had increased greatly since 1880.

1934-1969

In the early part of this period (up to 1950) this species was a quite common resident but was restricted in distribution to the areas away from fenland. By 1960 there had been a marked decrease and it had become uncommon; this was considered to be due to both the wide use of insecticides and the loss of preferred grassland habitat. In 1969 there was a sign of a recovery.

Present Status

A thinly distributed resident.

Found in summer only south of Ely. In 1970 it was seen in only around five parishes but by 1975 and subsequently has been noted in up to 20. The Cambridge Bird Club investigation (Bircham 1980) confirmed the distribution, but there was a single winter record in Ely and one in the far north of the county at Wisbech.

Breeding Status

The investigation gave a maximum of 21 locations where breeding might occur (present in summer) and a minimum of eight (breeding proved) with distribution strongest in the west of the county. This represents a considerable increase from the time of the *Breeding Atlas* (Sharrock) when breeding was confirmed in only two squares in the county.

BIRCHAM P.M.M. The status of woodpeckers in Cambridgeshire. *Camb. Bird Club Report* 54. 1980.

Great Spotted Woodpecker   *Dendrocopus major*

Pre 1934

Evans stated that it 'occurs in places but nowhere common'. Lack described it as resident, common in Cambridge, elsewhere local but most common in the west.

1934-1969

In 1944 it was described as being well distributed in 'suitable localities'. In 1948 it was considered 'commoner than the green'. There was then little mention until 1967 when the status summary considered it to be thinly distributed and local in fenland.

Present Status

Resident.
Well distributed in the area away from the fenland. Although in 1970 it was discovered in only nine localities, by 1975, and subsequently, this had risen to 25. In Cambridge there is a considerable population and many records are received each year from all areas of the city.

Breeding Status

The 1980 status (Bircham) showed a maximum of 33 (possible) localities and a minimum of 17 (proven). At some of the 17 more than one pair was present, e.g. Wicken Fen (2-3), Hayley Wood (2-3) and Sawston Hall Woods (2-3). Distribution of breeding pairs appears to be up to, and including, the southern fringes of fenland with a pair even on the Ouse Washes. The *Breeding Atlas* (Sharrock) shows the same general distribution with the addition of one or two sites in the north of the county where breeding was recorded.

Ringing Results

A single bird ringed at Great Raveley (Hunts) in late September 1981 was found dead at Six Mile Bottom in late August 1983 (42 kilometres distant).

BIRCHAM P.M.M. The status of woodpeckers in Cambridgeshire. *Camb. Bird Club Report* 54. 1980.

Lesser Spotted Woodpecker   *Dendrocopus minor*

Pre 1934

Evans stated of this species that it 'occurs in places but is nowhere common' and Lack considered it to be the scarcest of all the tree-climbing species with but a few pairs breeding. He added that it was possibly more common then than was supposed.

1934-1969

The status of this, the smallest of the three woodpeckers, is more confusing than for the other woodpeckers since the number of records per annum varied considerably. It is not possible to know whether this was due to changes in population or the fact that observers overlooked its presence. In the 1940s there was an increase in the number of records; however, by 1956 it was described as 'very rare away from Cambridge city'. In 1961 it was commoner, yet in 1964 a decrease was noted.

Present Status

Resident.

Widely reported away from the fenland area. A rapid increase in observations has led to it becoming the commonest of its genus in the county. Recorded at between 20 and 40 sites over the last ten years with a multitude of wintering records from the southern half of the county.

Breeding Status

The 1980 status (Bircham) showed it to be breeding at a maximum of 36 and a minimum of 18 sites, with more than one pair at some, and a distribution centred on Cambridge and the surrounding districts. The records received in 1983 suggested a possible total population of up to 50 pairs; certainly it is commonly seen throughout the breeding season and the population in Cambridge alone must be substantial. At suitable sites several pairs breed together.

BIRCHAM P.M.M. The status of woodpeckers in Cambridgeshire. *Camb. Bird Club Report* 54. 1980.

Short-Toed Lark   *Calandrella brachydactyla*

An extremely rare vagrant.
One record only:

A single bird caught by a bird catcher near Cambridge in November 1882 was identified by Alfred Newton. [Lack]

[Crested Lark   *Galerida cristata*

A record of a bird taken from a nest in 1881 at Histon was not accepted by Lack.]

Woodlark *Lullula arborea*

Pre 1934

Lack reported that it bred in Chippenham Park (report of a keeper) but added that it occurred nowhere else in the county.

1934-Present Status

A rare visitor.
Fourteen records since Lack, including a breeding pair, are all listed below:
Further records from Chippenham and Kennett suggest that a small breeding colony existed there up to 1957, being an overspill from nearby Breckland colonies.

 1 One at Milton, 22 April 1942.
 2 Six at Chippenham Park, 15 May 1948.
 3 Two singing between Chippenham and Isleham, 17 June 1948.
 4 One at Chippenham, 10 June 1951.
 5 One flying west over Cambridge, 24 October 1951.
6/7 Two separate individuals on passage at Coploe Hill, Ickleton, 8-9 August and 5 September 1954.
 8 A pair (may have bred) at Kennett Heath during 1956.
 9 Two at Royston sewage farm, 26 February 1956.
10 A pair bred at Kennett in 1957.
11 One singing on the Ouse Washes, at Pymore Bridge on 31 March 1963, was considered to be a passage bird.
12 One at Wicken Fen, 4 August 1972.
13 One at Quy Fen, 28 April 1974.
14 One with Skylarks, Foxton, 31 August 1980.

Skylark *Alauda arvensis*

Pre 1934

Evans stated that this species was 'very common' and Lack described it thus 'in fact one of our commonest species' found in fields all over the county, and present in large flocks in winter. He particularly noted that it was heard on migration at night over the city.

1934-1969

Unchanged from Lack. The main point of interest centred around the large-scale weather movements during winter months with up to 2000 per hour. In the more severe conditions birds were heading in a westward direction, in front of hard weather spreading from the Continent.

Present Status

A very common and widely distributed resident.

Probably one of the few species not adversely affected by the changes in agricultural practice in the county. Autumn passage and hard-weather movements are regularly noted, most commonly in January, often of a similar scale to those above. Counts are hard to come by but 400 were noted in a field of oilseed rape near Hauxton in January 1984 showing how the introduction of this crop has provided a new habitat for this and other species.

Breeding Status

A common and widely distributed breeding species on arable land, as the *Breeding Atlas* (Sharrock) shows. Counts on the Ouse and Nene Washes

Figure 42. Skylark – the number of breeding pairs counted at Barton Farm, Coton common bird census site.

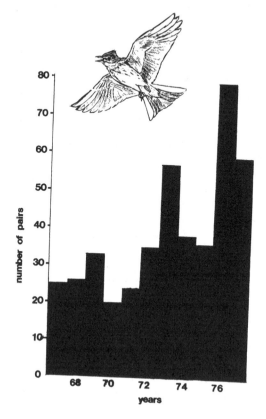

in 1982 revealed 339 and 360 singing males respectively. As a species likely to benefit from 'prairie' farming conditions increases on two Common Bird Census plots from 18 pairs at Soham and 19 pairs at Knapwell in 1963 to 24 and 35 pairs respectively by 1967 indicate the level of success (Bibby 1970). At Barton Farm, Coton, the hedge removal in 1969 led to a dramatic increase in the breeding population as shown in Figure 42. (Bucknell 1977).

BIBBY C.J. The Common Bird Census. *Camb. Bird Club Report* 44. 1970.

Sand Martin   *Riparia riparia*

Pre 1934
Evans considered it a 'common summer visitor' and Lack stated that it was a summer resident which bred locally at the few suitable sites. He added that it was very common on autumn migration which usually began in the first week of August and lasted through until the end of September.

1934-1969
Most of the records from this period refer to gatherings, usually on or around Cambridge sewage farm where up to 2500 occurred on passage in autumn. In 1947 breeding colonies at Chippenham (80 pairs) and Shippea Hill (20 pairs) were mentioned and by 1949 there were records from the 'Backs' of up to eight pairs and from Mepal pits where up to 500 pairs were reported. By 1959 colony counts included 200 pairs at Milton GP and 250 pairs at Chippenham. Two years later (1961) there was a roost on the Ouse Washes estimated at 1 000 000 birds! Although this was an annual event the numbers were generally considerably lower with the exception of 1968 when there was an estimate of 2 000 000 during August!

Present Status
A regular but local summer visitor.
Present from late March to early October, in line with the national trend most Cambridgeshire observers report a deterioration in the status of this species at the time of writing. There are only one or two breeding colonies at gravel pits and along the banks of the River Cam in Cambridge. Roost numbers have varied between hundreds and thousands with 70 000 at Ely beet factory in 1983 and 20 000 on the Ouse Washes in the same year. The timing and siting of these roosts is variable.

Breeding Status

This species breeds in most years although the numbers and sites vary. In 1984 there were 60 pairs at Waterbeach GP, six pairs at Whittlesford GP, 'good numbers' at Whittlesey, and the usual few pairs along the Cam.

Ringing Results

Most of those birds that are ringed in the county are caught at autumn roosts and their origins are revealed in the map (Fig.43) showing that most of these birds are controls journeying south presumably on the first stage of their migration. There is no evidence that the time spent at each site is long and very few are caught on more than one occasion no matter how often the site is ringed.

Earliest date : 5 March (1961 Earith)
Latest date : 17 November (1968)

Swallow   *Hirundo rustica*

Pre 1934

Described by Evans as an 'abundant summer visitor'. Lack considered it to be a summer resident nesting in suitable habitat but not abundant! Con-

Figure 43. Sand Martin and Swallow – British ringing recoveries of birds ringed in Cambridgeshire with arrows to show the direction of movement for birds ringed and recaptured in the same year.

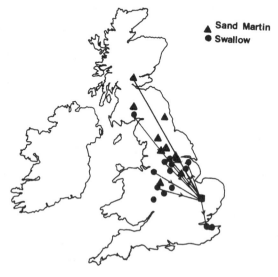

▲ Sand Martin
● Swallow

siderable spring and autumn migration was observed particularly at Cambridge sewage farm.

1934-Present Status
An abundant summer resident.
Records of large gatherings are usually up to 500 and associated either with freak feeding conditions or migrational assemblies. Roosts of large numbers (up to 20 000) have occurred at Ely beet factory, Wicken Fen and on the Ouse Washes; these are often associated with other hirundines. Most areas of reedbed, however small, attract roosting birds in autumn.

Breeding Status
Widespread, the distribution being dependent on available nest sites. An amazing record in 1964 described a nest built in a lorry at RAF Ely hospital. The adults fed the young to fledging despite the fact that the lorry was in use and moved about the hospital grounds and on one occasion made a trip to Wisbech and back being therefore absent for four and a half hours!

Ringing Results
There have been many recoveries since birds of this species are caught in large numbers at their roosting sites. The map (Fig.43) shows that most of these birds, like the Sand Martin, originate in the north of England. Three April recoveries in MOROCCO of birds ringed at: Milton in 1962, Wicken Fen in September 1970 and September 1974, and another in ALGERIA of one ringed at Wicken Fen in September 1980 give an indication of the return route taken by these birds. The only recovery in winter quarters is a bird ringed at Wicken Fen in August 1962 which was reported from Orange Free State, SOUTH AFRICA, in January 1963.
Earliest date: 19 March (1977 Ouse Washes)
Latest date: 3 December (1953 Ely Cathedral where a family was raised in November!)

Red-Rumped Swallow   *Hirundo daurica*

An extremely rare vagrant.
One record only:
A single bird was reported on the Ouse Washes on 6 May 1984.

House Martin   *Delichon urbica*

Pre 1934

Evans described this species as a 'plentiful summer visitor' and Lack stated that it was a summer resident and a more plentiful breeding species than the Swallow but less abundant than in other parts of England. He added that it was not so common on migration. Evans pointed out an obvious decline in Cambridgeshire during the period 1890-1930 that was not reflected in other counties.

1934-Present Status

A common summer resident.

Present from mid April to late October. In some years there have been gatherings several hundred strong, most often during October as birds assemble for departure. House Martins are often to be found at the roosts of other hirundines but usually only as a small proportion of the total.

Breeding Status

In the *Breeding Atlas* (Sharrock) it is shown to be a common breeding species with colonies associated with buildings all around the county.

Ringing Results

A bird ringed at Ingatestone (Essex), in early August 1969, was caught at Wicken Fen in May 1973; presumably this bird was migrating out when first caught and in when controlled.

Earliest date: 10 March (1983 Hinxton)

Latest date: 27 November (1940)

Richard's Pipit   *Anthus novaeseelandiae*

An extremely rare vagrant. Three records, all listed below:

1 One at Wilburton on 20 November 1952 was caught in a shed and examined, before being released the following day.
2 One on the Ouse Washes, 29 October 1967.
3 One at Waterbeach GP, 22 November 1976.

Tawny Pipit   *Anthus campestris*

An extremely rare vagrant. Two records listed below:

1 One at Grantchester, 7 September 1962.
2 One at Kennett GP, 11-12 June 1972.

Tree Pipit   *Anthus trivialis*

Pre 1934

Evans described this species as a 'local summer visitor, nowhere plentiful'. Lack went a stage further and stated that it was a not uncommon summer visitor, found chiefly in the west and south of the county. Farren told Lack that it had been common until about 1906 when numbers fell.

1934-1969

Throughout this period there were several records each year and breeding was regularly reported from Chippenham (abundant 1948) and sporadically elsewhere as at Little Eversden (1-2 pairs 1948 and 1949), the Gog Magog Hills (up to 4 pairs 1960), Ditton Park Wood (4-6 pairs in the 1960s), and the Devil's Dyke (one pair 1969). There were scattered records of passage birds in both spring and autumn; several were noted at Cambridge sewage farm and Milton GP.

Present Status

An unusual passage visitor in both spring and autumn, and an irregular breeding species.
Records are confined to the area away from the fenland. Singing males are recorded but further evidence of breeding is not always forthcoming, which suggests that these may be passage birds and it is likely that passage itself is far larger than that which is reported. It may prove to be a lack of observation but it seems probable that there has been a change in the status of this species and that it is in decline in the county.

Breeding Status

The only recently proved breeding record was of a pair on the Gog Magog Hills in 1978, and although a pair probably nested at Ditton Park Wood in 1970 the former record was the first confirmed since 1969 when a pair nested on the Devil's Dyke.
Earliest date: 13 April (1949 Over)
Latest date: 3 November (1952 Cambridge sewage farm)

Meadow Pipit   *Anthus pratensis*

Pre 1934

Evans described it as 'not uncommon, somewhat local'. Lack, however, stated that it was common throughout the county breeding in meadows and marshland. He considered it to be scarcer in winter and scattered in distribution.

1934-Present Status

A common resident.

Found throughout the county, with numbers increasing during peak passage periods. The loss of grassland through this period has led to a decline in the breeding population but it is probably more common in winter than in Lack's time and is subject to dramatic movements in hard weather.

Cuckoo at a Meadow Pipit's nest on the Devil's Ditch.

Breeding Status

The breeding distribution in the *Breeding Atlas* (Sharrock) is rather irregular. The south Cambridgeshire population has been especially low since the 1950s, with scattered sites holding only a few pairs. In fenland it is more widely recorded but the only substantial colonies remain on the washlands. Counts have been made regularly on the Ouse Washes showing a population of between 100 and 500 pairs with around 250 in most years (not all in Cambridgeshire). On the Nene Washes counts have been: 150 pairs in 1982, 46 in 1983 and 280 in 1984.

Rock Pipit   *Anthus spinoletta*

Pre 1934

While not mentioned by Evans, Lack gave eight or nine records of birds at Cambridge sewage farm which he stated were mostly of 'Water' Pipit (*A.s.spinoletta*). The first was in October 1915, the remainder appearing each winter from 1930. Although these were the only records, Lack considered that this species was a regular winter visitor that had been overlooked.

1934-1969

Up to the mid 1950s this species was recorded annually at Cambridge sewage farm between October and April. Numbers: usually single birds but maximum of five. Both 'Water' and the Scandinavian form (*A.s.littoralis*) were seen with some regularity although rarely more than one or two per annum. Into the 1960s this bird was noted at other sites particularly on the Ouse Washes and the developing gravel pits: Milton, Landbeach, Waterbeach and Fen Drayton.

Present Status

A regular, but uncommon, passage migrant and occasional winter visitor. Recorded between October and April but seen most commonly in March. Favoured sites include Ely beet factory, Ouse and Nene Washes and the various gravel pits. Numbers : usually 1-2 but exceptionally 14 on the Ouse Washes in March 1975. One or two 'Water' and Scandinavian birds continue to be seen somewhat irregularly.

Earliest date: 1 October (1965 Landbeach GP)

Latest date: 18 April (1934 Cambridge sewage farm)

Yellow Wagtail   *Motacilla flava*

Pre 1934

Evans described this species as 'a very common summer visitor'. Lack stated that it was a fairly common summer visitor breeding in marshes and meadowlands especially in fen districts and river valleys.

1934-Present Status

A regular summer resident.

Over the post-Lack period it has been regularly reported between early April and early October. Numbers are swollen in spring with passage birds so that large congregations can occur, e.g. 250 on the Ouse Washes in April 1963. Favoured sites include The Ouse, Nene and Cam Washes, Wicken Fen, Ely beet factory and the gravel pits. The common form is *M.f.flavissima*. Among other races 'Grey headed' (*M.f.thunbergi*) and 'Sykes' (*M.f.beema*) have occurred but only the 'Blue headed' (*M.f.flava*) is noted at all regularly.

Breeding Status

A regular breeding species in small numbers mainly in the arable areas of fenland and central Cambridgeshire with pairs at one or two gravel pits in the south. There are good populations on the various washlands, although numbers depend on the water level. On the Ouse Washes numbers counted include an estimate of 20-30 pairs per square mile in 1953, and counts of c.100 pairs in 1973, 77 in 1974, 262 pairs in 1972, 293 pairs in 1977, 181 in 1978, 145 in 1980 and 362 in 1982. Sorensen (1974) pointed out that population trends on the Ouse Washes were in contrast to those elsewhere (nationally) due to management techniques designed to provide extra habitat for this species. On the Nene Washes a smaller population of 20-60 pairs is increasing and on the Cam Washes up to 12 pairs are usually present. One to three pairs can be found at Wicken Fen, Ely beet factory and the various gravel pits.

Earliest date: 26 March (1955 Six Mile Bottom)

Latest date: 27 December (1968 Ouse Washes)

SORENSEN Jeremy. Population trends of some small passerines on the Ouse Washes. *Camb. Bird Club Report* 48. 1974.

Grey Wagtail   *Motacilla cinerea*

Pre 1934

Evans considered this species to be a 'rare winter visitor' but by Lack's time it had become regular arriving in the first week of October and leaving in March. Numbers were small, rarely more than six together but usually single birds, seen along the banks of the River Cam, at Cambridge sewage farm or other areas of standing water, and rarely in the fens. H.G. Alexander reported to Lack that he had heard one singing near Grantchester in early May 1911 but no other summer records had been received.

1934-1969

By the greatest coincidence the first breeding record was received in the year Lack's book was published: 1934 at Hildersham. By 1938 there was a pair at Linton as well and although in the main records referred to winter visitors one or two pairs bred in that area of the county in most years. In 1953 four sites were found including the original but also at Lode, Swaffham Bulbeck and Bottisham. During the 1960s breeding records were spasmodic with the most regularly used site being at Lode although other sites may well have been overlooked. Towards the end of this period most records were of visitors between mid September and mid April.

Present Status

An uncommon resident, with a small breeding population and passage and winter visitors.
Numbers: 1-3 out of the breeding season, generally distributed away from the fenland area. Favoured sites include the upper reaches of the River Cam (Granta), Cambridge, and Ely beet factory, with occasional records from the various gravel pits.

Breeding Status

Although not breeding at the beginning of this period nesting was discovered at Cambridge, Shelford, Linton, Hauxton, Ickleton and Grantchester at least once over the whole period with several of these locations used more than once. Total breeding numbers therefore have been about 1-3 pairs per annum. The *Breeding Atlas* (Sharrock) shows this species to have a strong breeding distribution in the western half of the country but to be absent from large parts of the eastern half.

Pied Wagtail   *Motacilla alba*

Pre 1934

Evans described it as 'abundant'. The 'White' wagtail (*M.a.alba*) he described as a regular spring and autumn visitor and he stated that pairs nested in ivy on Downing and King's Colleges. Lack stated that it was scarce in winter but commoner in summer and more widely distributed. Roosts were noted at Adams Road Bird Sanctuary in Cambridge and by King's College gate.

1934-Present Status

A common resident.

Most records of interest are of roosts. Numbers up to 500 can be seen together but smaller gatherings are more usual. Favoured roost sites include Fowlmere 100-500 maxima, Milton up to 300, Fulbourn Fen 40-120, Cambridge sewage farm 100-250, and Chippenham and Wicken Fens up to 100. 'White' wagtails are recorded on passage in most years particularly in spring.

Breeding Status

Thinly distributed and shown in the *Breeding Atlas* (Sharrock) to be absent from one or two areas of the county. However, it nests regularly at scattered sites across the county, as at sewage farms, factory sites, etc., and is an opportunist occurring wherever suitable sites become available.

Ringing Results

Birds caught at roosts, particularly in Milton, have yielded a little information. A bird ringed in July 1971 at Fowlmere was recovered at Much Hadham (Herts) in March 1976. One ringed at Abberton (Essex) in September 1956 was found in Cambridge in December 1957. One ringed in Cambridge in March 1961 was found in Wishaw (Lanarkshire) in June 1961. Finally, two birds ringed at a roost in Milton in September 1959 were found in SPAIN at Huelva and at Arcos, Cadiz, in the October and November respectively of that year.

Waxwing   *Bombycilla garrulus*

Pre 1934

Evans stated of this bird that it 'occurs at intervals in winter, singly or in flocks'. Lack noted small parties generally in January or February in the following years: 1825, 1850, 1873, 1893-95, 1903, 1914 and 1921. The largest party was 20 near Wisbech in 1850.

1934-1969

In the early part of this period there were records only in: 1934, 1944, 1947, 1949, 1951 and 1952. In the first four of these years there were many records but only 1-3 in the others. From 1957 to 1966 there were about 1-5 records each year, with the exception of the winter 1965/1966 when there was an autumn irruption on the Continent leading to many records throughout the winter. These were analysed by Cornwallis and Townsend (1968). Numbers: 1-10 in general although exceptionally in the irruption winter there were 40 east of Peterborough and several smaller flocks of up to 15. Seen all around the county but more often recorded in Cambridge probably due to the interest of observers and the relatively conspicuous nature of the bird.

Present Status

An irregular winter visitor.
Recorded between November and March in small numbers (1-3) around the county. There were a number of sightings in both 1970 and 1971 but since then there have only been ten records.
Earliest date: 5 November (1961 Milton)
Latest date: 4 April (1951 Fordham)

CORNWALLIS R.K. and TOWNSEND A.D. Waxwings in Britain and Europe during 1965/66. *Brit. Birds* 61. 1968.

Dipper   *Cinclus cinclus*

An extremely rare vagrant. One record only:
A a single bird (on a single occasion) in the winter 1949-50 by the stream between Snailwell and Fordham, the date was not noted but the description was so typical of this species that the record was accepted.

Wren   *Troglodytes troglodytes*

Pre 1934

Described by Evans as 'common' and by Lack as a common and universally distributed resident found not only in woods and gardens but also reedbeds.

1934-Present Status

This species remains an abundant and widespread resident breeding in every suitable location.

Roosts are common in winter although the numbers are usually small. At Wicken Fen during the harsh winter of 1947 up to 70 birds of this species were seen 'darting into sheds, hen coops, etc. at dusk'. This species is also found at some wetland sites where any scrub exists as is shown by the fact that an average of 75-80 are ringed each year at Wicken Fen, and in 1984 209 were ringed in comparison with 155 Sedge Warbler.

Breeding Status

Widespread and abundant. Shown in the *Breeding Atlas* (Sharrock) to be in every 10 km square. Harper (1982) stated that in the Botanic Garden males were sometimes polygamous and that at least 12 were breeding in 1981. Conder (1975) gave a density of between 20 and 50 pairs at Hayley Wood.

Ringing Results

Out of the hundreds of Wren ringed in the county the only non-local recoveries are of birds ringed in September at Wicken Fen. Of two in 1973 one was recovered at Dartford (Kent), and the other on Jersey (Channel Islands); and one ringed in 1980 was found at Bexleyheath, London, in March 1982. The population in general is highly sedentary.

Dunnock   *Prunella modularis*

Pre 1934

Described by Evans as 'common' and by Lack as fairly numerous but not as common as the Robin.

1934-Present Status

An abundant and widely distributed resident.

Breeding Status

Widespread and abundant. The *Breeding Atlas* shows it to be present in each 10 km square in the county (Sharrock). This species has been the subject of a behavioural study in the Botanic Garden and the number of nesting birds is known to be of the order of 20-30 pairs. (N.B. Davies pers. comm.). In Hayley Wood Conder (1975) estimated a population of up to

10 pairs and the Coton census (Bucknell 1977) counted between 11 and 28.

Ringing Results

Like the Wren, this species has been shown through ringing to be highly sedentary. The only long-distance recovery is of a bird ringed on Fair Isle in April 1962 which was found dead at March six days later.

Alpine Accentor   *Prunella colaris*

An extremely rare vagrant. Two records listed below:

1 Two at King's College on 22 November 1822 were the first record for Great Britain. The female was subsequently shot. [Lack]
2 A pair on the 'Backs' near Clare College, 30 April 1931. [Lack]

Robin   *Erithacus rubecula*

Pre 1934

Described by Evans as 'common' and by Lack as a common and widely distributed resident.

1934-Present Status

An abundant and widely distributed resident in all parts of the county although obviously more sparsely in open fenland.

Breeding Status

An abundant and widespread breeding species shown in the *Breeding Atlas* to be present in every 10 km square (Sharrock). In the Botanic Garden Harper (1982) quoted a population of 15-20 pairs and in Hayley Wood Conder (1975) estimated 20-50 pairs. On the Coton farmland site between 4 and 16 pairs were counted (Bucknell 1977).

Ringing Results

Of all the birds of this species ringed in the county only a few have displayed any interesting movement. The only foreign recovery is a bird ringed in April 1982 at Wicken Fen that was killed by a cat at Kalmar, SWEDEN, in April 1983. Two birds moved to King's Lynn (Norfolk), and Skendlesby (Lincs); and one arrived at Wicken Fen in October 1976 having been ringed at Sheringham (Norfolk) a month earlier (a migrant?). Finally, a bird found at Wantage (Berks), in March 1939, had been ringed in Cambridge in January 1937.

Thrush Nightingale   *Luscinia luscinia*

An extremely rare vagrant.
One record only:
A single bird was seen on the Ouse Washes, 2-3 June 1984.

Nightingale   *Luscinia megarhynchos*

Pre 1934

Evans considered this species to be 'an abundant summer visitor in most parts'. Lack stated that it was common throughout woods in the south of the county but less abundant in fen districts. Ticehurst, in his review of Lack's Birds of Cambridgeshire, suggested that along Trumpington Road, Cambridge, there were more birds of this species per mile than any other stretch of road in England!

1934-1969

A common summer resident found in several areas in the county with a strong breeding population particularly in Madingley Wood where up to nine pairs have been reported. Towards the end of this period, in the late 1960s, there were signs of a decreasing population particularly in 1968 when there were few occupied fenland sites.

Present Status

A summer resident.
Present from mid April to late August at traditional sites. Its status has varied considerably over the years 1970-86 and it is probably no longer as common as in the previous period.

Breeding Status

In 1976, the year of the first BTO status survey, only three pairs were located and this species was considered by Hudson (1979) to be in a national decline similar to that shown in the early stages by both the Wryneck and the Red-backed Shrike. Since the dreadful year of 1976, however, numbers have recovered and by 1980 a possible 19 pairs were counted, the highest number recorded for many years. Over the years since 1980 10-15 pairs have been reported with strong representation at Wicken Fen (up to 4 pairs) and the Kennett-Chippenham area (up to 4 pairs). One or two pairs nest at the other sites which include Fen Drayton GP, Fowlmere, Fleam Dyke, Fordham and Hayley Woods, Landbeach GP,

Milton GP and the Gog Magog Hills, although all these sites are not always occupied. Davis (1982) in his review of the second BTO census stated that 1980 was a very good year nationally. In conclusion it should be pointed out that Cambridgeshire being both short of suitable woodland and on the northern edge of the breeding range of this species is probably especially subject to fluctuations of the kind detailed above.

Earliest date: 3 April (1947 Botanic Garden, Cambridge)
Latest date: 30 August (1952)

DAVIS Peter G. Nightingales in Britain in 1980. *Bird Study* 29. 1982.
HUDSON Robert. Nightingales in Britain in 1976. *Bird Study* 26. 1979.
TICEHURST N.F. Review of 'The Birds of Cambridgeshire'. *Brit. Birds* 28. 1935.

Bluethroat   *Luscinia svecica*

An extremely rare vagrant.
Three records, listed below:
[A record of two boys seeing one at Chatteris in May 1873 is disregarded by Lack on the grounds of insufficient evidence.]

 1 An immature seen at Cambridge sewage farm on 9 August 1947.

[This record was considered suspect in the 1947 *Camb. Bird Club Report* but one of the observers (R.S.R. Fitter pers. comm.) confirms that there was no doubt as to the identification. The date was perhaps considered early but in view of record 3 this no longer seems a relevant consideration.]

 2 One Red-spotted (*L.s.svecica*) was seen in Rustat Road, Cambridge, on 9 August 1979.
 3 One White-spotted (*L.s.cyanecula*) at Wicken Fen, 30 March 1980.

Black Redstart   *Phoenicurus ochruros*

Pre 1934

Lack quotes three records, one sent to Farren from near Cambridge in the 1890s, one at St John's College in November 1915, and one on Corpus Christi College chapel in October 1927.

1934-1969

In the immediate post-Lack period there was a spate of breeding records: a probable in 1936, a pair in 1937, 1938 and 1940 and in 1942 at least three pairs nested on University property in Cambridge, (the Chemistry Laboratory, St John's College and Queens' College), together with singing males at Corpus Christi, in the Cauis/Clare area and at Magdalene. In addition an immature male sang both inside and outside Ely cathedral! All this activity was attributed to a westward movement through Europe. In 1943 several pairs may have nested and in the following years there were records in May around the centre of Cambridge but no breeding. After some irregular records a pair bred at Ely beet factory in 1958 but it was not until 1969 that there was another such record and that was of a male summering around St John's College. Otherwise records were sparse during the latter half of this period.

Present Status

A passage migrant, winter visitor and occasional breeding species. Numbers: usually single birds, which can occur in any month of the year although records in high summer are unusual unless breeding is involved. Locations from which it is reported are widespread.

Breeding Status

There have been a few breeding records since 1969. A pair bred in 1970 in Cambridge, at the New Addenbrooke's Hospital buildings, which were under construction. In 1980 a pair bred at Wisbech Hospital and in the 1980s pairs have been found in most years mainly in or near Cambridge centred on the college gardens.

In the 1930s and 1940s, the Black Redstart nested frequently near the Cambridge Colleges.

Redstart *Phoenicurus phoenicurus*

Pre 1934

Jenyns described this species as very abundant, and it remained abundant until about 1885-90 when there was a notable decline according to Evans and Farren. [Lack] By the turn of the century it was reduced to the status of a local breeding species and even though several pairs were nesting in the vicinity of Cambridge up to 1915, from 1927 to 1933 only two or three sites were reported for the whole county.

1934-1969

In the late 1930s breeding records continued to be received, mostly from Hauxton where up to three pairs nested. The last positive record was in 1938, but there were later records of singing males. By 1940 this species had declined to be a passage visitor with one or two records of birds in summer and the decline culminated in there being no records at all in 1953. In the years that followed there were regular sightings in both spring and autumn, usually of single birds (maximum two) at many sites around the county. In 1966 a pair bred at Bourn and in 1968 breeding was reported from one or two sites.

Present Status

A regular spring and autumn passage migrant with some irregular breeding.

Reports come from all around the county and usually involve 1-2 birds stopping briefly at a site only for a few days. Generally records fall between, 10 April to 10 May, and 1 August to 20 October.

Breeding Status

A recent record of three pairs at a site in the west of the county in 1974 is the only report in this category.

Earliest date: 4 April (1985 Wicken Fen)

Latest date: 29 October (1976 Whittlesford)

Whinchat *Saxicola rubetra*

Pre 1934

Evans considered it a fairly common summer visitor. Lack stated that it was sparsely distributed and found mainly on wasteland, heaths and dykes, nowhere common and absent from many areas.

1934-1969

Throughout the period until the late 1960s there was a very regular breeding population which was summarised in 1955 as follows: centred on the Ouse and Nene Washes with a density of 2-4 pairs per square mile, present in the area of Fulbourn Fen, about nine pairs in 1955 and found at twelve other fenland sites. In 1966 the first warning sign of impending decline was seen with fewer reports and by 1969 it bred only at Milton. The reason for this decline remains rather obscure.

Present Status

Mainly a spring and autumn passage migrant with some sporadic breeding records.

Spring passage is mainly between 20 April and 31 May while autumn is between 1 August and 20 October. Numbers: usually 1-2 but exceptionally 5-6 at a site. Sites such as the Ouse and Nene Washes and Fowlmere are regularly visited but individuals may occur almost anywhere in the county.

Breeding Status

Nesting continued with records from near Hayley Wood and Fowlmere watercress beds in 1970, on the Ouse Washes (up to 5 pairs) and Graveley in 1971, on the Nene Washes (2 pairs) in 1972 and a single pair there in 1973. In 1974 none were found nesting and with the possible exception of a pair at Milton in 1981 none have bred since. The *Breeding Atlas* (Sharrock) shows the distribution of this species to be predominantly in the west of the country and it may be that there has been a general decline in the east.

Earliest date: 18 April (1964 Ouse washes)
Latest date: 5 December (1976 Whittlesford where an individual was seen following the plough in hard weather!)

Stonechat   *Saxicola torquata*

Pre 1934

Jenyns noted this species to be very common in the fens, but Evans stated that it was 'very local' by the end of the nineteenth century. Lack described it as breeding at Wicken and neighbouring fens and some other waste areas in the county. He added that it was commoner in winter. The general decrease from Jenyns' time probably came about due to increased cultivation about 1880.

1934-1969

Following Lack's publication there were no more proven breeding records although some were suspected in the immediate period up to the late 1940s. This species then became an uncommon passage migrant and/or winter visitor with an average of five or six records each year. Numbers: mostly singles with sometimes two birds or exceptionally three at any one site. In the mid 1960s the number of records dropped to barely one or two per annum.

Present Status

A regular visitor out of the breeding season.
The number of records each year suggests a recovered status from that at the end of the 1960s. On the Ouse and Nene Washes in recent years 1-5 birds have been present from October to April and long-stay birds have been noted at other sites around the county more often north of Cambridge. Numbers: usually only 1-2 per site. There have been no further breeding records.

[Isabelline Wheatear   *Oenanthe isabellina*

E.A.R. Ennion reported a pair at Burwell Fen in April 1929, and although the description was considered good, the record was rejected 'without a specimen to support it'.]

Wheatear   *Oenanthe oenanthe*

Pre 1934

Jenyns noted it breeding at the Devil's Dyke and Evans stated that it was occasionally observed on migration and sometimes bred. He found nests on the Devil's Dyke in the 1880s and 1890s. Lack stated that it bred around Chippenham, and a nest was found near Cambridge in 1927. Elsewhere in the county it was to be seen only on migration.

1934-1969

Generally confined to a passage migrant status in both spring and autumn. Probable/possible breeding occurred in 1944 at Wicken Fen, near Isleham in 1945 and at Chippenham and Shippea Hill in 1948, all these sites being on the eastern edge of the county and thus nearest to the Breckland population. Numbers on passage were generally small but exceptionally

35 were counted in the Melbourn area in late spring of 1954 and 18 in a 250-acre field in 1953.

Present Status

A regular spring and autumn passage migrant.
Can be seen anywhere in the county, and is more common in spring than autumn. Generally noted as follows: end of March to the end of May (often in two waves) and from the end of July to the middle of October. Numbers: usually 1-2 but sometimes more and exceptionally up to 50 as on the Gog Magog Hills in April 1980. Birds of the Greenland race (*O.o.leucorrhoa*) frequently pass through, usually in late May.

Breeding Status

A pair raised young at Kennett in 1970 in a Breckland-like situation, but there have been no records since.
Earliest date: 7 March (1977 Ouse Washes)
Latest date: 24 November (1957 Melbourn)

Ring Ouzel   *Turdus torquatus*

Pre 1934

Evans stated that it 'occurs rarely on autumn migration and even less commonly in spring'. Lack quoted around ten records and described it as a passage migrant noted in both spring and autumn.

1934-1969

There was a maximum of 21 records in this period, 14 on spring and seven on autumn migration. Apart from one record of two at Wicken Fen in October 1965 all were of single birds.

Present Status

An uncommon, but regular, passage migrant, mainly in spring.
There have been 39 records since 1970 and they have come from all around the county, suggesting either greater numbers or increased observations. Numbers: nearly all single birds, though exceptionally five birds stayed at Wandlebury for three weeks in April 1974, and five and four were seen on the Nene and Ouse Washes respectively in May 1984. The periods of passage are mid March to mid May and mid September to late October.

Ringing Results
A single bird ringed at Fulbourn Fen in November 1964 was found at
Blegier, Basses Alps, FRANCE, in December 1969.
Earliest date: 13 March (1951 Kneesworth)
Latest date: 27 October (1962 Fulbourn Fen)

Blackbird   *Turdus merula*

Pre 1934
Evans described this species as 'common' and Lack reported that it was
exceptionally abundant and universally distributed.

1934-Present Status
An extremely abundant and widely distributed resident.
Some immigration takes place in autumn, usually around late October,
but many of these birds probably move on south or west after a short stay.

Breeding Status
Abundant and widespread! On the Coton census site the number of

Figure 44. Blackbird and Song Thrush – foreign recoveries of birds ringed in
Cambridgeshire.

o Song Thrush
● Blackbird

territories between 1967 and 1977 varied from 13 to 36 (Bucknell 1977) and in the Botanic Garden Harper (1982) estimated a population of about 80 pairs.

Ringing Results

There are a large number of foreign recoveries showing the origins of the winter visitors and these are displayed in Figure 44. All these birds were found or ringed in Cambridgeshire during the period October to February and the monthly distribution is as follows: October 5, November 4, December 5, January 6 and February 3. The only bird found in Cambridgeshire in summer was a bird ringed in DENMARK on 1 October 1983 which was found at Wisbech on 3 June 1984. Bircham and Bibby (1978) analysed the recoveries and noted that contrary to the national picture none of the Cambridgeshire birds was from Norway. Onward movement was demonstrated by two birds ringed in November 1969 which were recovered in Oxford and Devon subsequently in the same winter.

BIRCHAM P.M.M. and BIBBY C.J. Movements of birds to and from Cambridgeshire. Thrushes. *Camb. Bird Club Report 52.* 1978.

Fieldfare   *Turdus pilaris*

Pre 1934

Evans described this species as a fairly abundant winter visitor. Lack agreed, with the addition that big flocks were seen throughout the county.

Figure 45. Fieldfare and Redwing – foreign recoveries of birds ringed in Cambridgeshire.

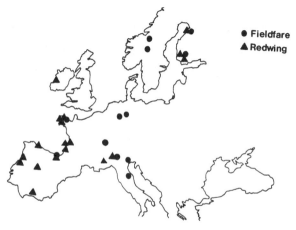

1934-Present Status

A common and widely distributed winter visitor.
Fieldfares generally arrive in the first week of October and leave in April.
Quite large gatherings of up to 1000 birds are seen in parts of the county,
particularly in autumn after arrival. Roosts of up to several thousands can
be found in suitable habitat at sites such as Fulbourn and Wicken Fens.
The timing of the main arrival in autumn is variable but usually there is a
considerable influx in the period between the middles of October and
November. Generally the numbers remain high for a month or so before
seeming to disperse with the onset of hard-weather conditions so often
prevalent in the new year. Milwright (1983), as a result of ringing large
numbers in orchards, has suggested that the actual individuals concerned
may change over the months.

Ringing Results

The foreign recoveries (Fig.45) show the origins of our winter visitors.
Bircham and Bibby (1978), using additional information from a roost in
Holywell which is just over the county boundary (Hunts), showed that
most birds came from Scandinavia, with one from the USSR and another
from Poland. Birds were subsequently found in other winters in FRANCE,
ITALY, PORTUGAL and SPAIN. Many of these birds were victims of
shooting.
Earliest date: 17 July (1972 Chettisham)
Latest date: 24 May (1980 Ouse Washes)
(A single bird was seen on the Ouse Washes on 1 July 1969 but due to the
date it is impossible to tell whether it was a early arrival or a late departure
or indeed if it left these shore at all.)

BIRCHAM P.M.M. and BIBBY C.J. Movements of birds to and from Cambridgeshire.
  Thrushes. *Camb. Bird Club Report* 52. 1978.
MILWRIGHT R.D.P. The Timing and duration of migratory waves of Redwing and
  Fieldfare in fenland orchards of mid Cambs and south-west Norfolk. *Wicken Fen Group
  Report* 12. 1983.

Song Thrush   *Turdus philomelos*

Pre 1934

Evans described this species as 'common'. Lack stated that it was a
common resident, less abundant than the Blackbird and was often heard
migrating over Cambridge in October and November.

1934-Present Status

An abundant and widespread resident.
Like the Blackbird, many individuals migrate from the Continent in autumn and this species, together with the Redwing, is a common sight feeding on the haws in October and November. Most of these birds seem to move further to the south as the really hard weather approaches.

Breeding Status

A widespread breeding species. In the Botanic Garden about 25 pairs nest (Harper 1982) while Conder (1975) estimated 5-20 pairs in Hayley Wood. On the Coton census site numbers varied from 6 to 15 pairs.

Ringing Results

Figure 44 shows the foreign recoveries for birds ringed or found in the county. Most of these are birds ringed one winter in Cambridgeshire and recovered during a subsequent winter in SPAIN or FRANCE. Two ringed at Wicken in October 1962 were found in FRANCE and PORTUGAL in the new year of the same winter. Very little information has emerged concerning the origins of these immigrants.

Redwing   *Turdus iliacus*

Pre 1934

Evans considered it to be a fairly abundant winter visitor and Lack stated that it was a common winter visitor, more numerous than the Fieldfare, with large numbers roosting at Adams Road Bird Sanctuary. It was also reported to be heard migrating over Cambridge in October and November.

1934-Present Status

A regular and widespread winter visitor.
Most birds arrive in early October and leave by late April. Some gatherings of up to 500 occur, particularly in urban areas such as playing fields and greens, and in larger gardens 10-20 may gather in hard weather particularly under shrubs. Roosts comprising all species of thrush often contain many of this species but exclusive roosts are unusual. Milwright (1983) found that in fenland orchards Redwing peaks occurred within the first part of the winter and that numbers were much reduced after the new year. He concluded that these birds had moved further south. However,

there is no evidence that all the winter visitors move on. In recent years there have been a few records of territorial singing in April and in 1984, for instance, a male held a territory at Wicken Fen for about ten days.

Ringing Results

Like the Song Thrush there is not much information about the origin of the migrants (two from FINLAND). A good number of recoveries in either the same or subsequent winters have been received, some in Britain, some abroad and these latter are displayed in Figure 45. Two birds ringed at Fulbourn Fen in October 1962 were found dead on the south coast within a couple of months of ringing, suggesting the sort of movement to be expected in such a harsh winter.

Earliest date: 24 September (1960 Milton)
Latest date: 3 June (1981 Gog Magog Hills)

MILWRIGHT R.D.P. The timing and duration of migratory waves of Redwing and Fieldfare in fenland orchards of mid Cambs and south-west Norfolk. *Wicken Fen Group Report* 12. 1983.

Mistle Thrush   *Turdus viscivorus*

Pre 1934

Evans described this species as 'plentiful' which is rather surprising since he rated the Song Thrush as only common. Lack stated that it was a widespread resident.

1934-Present Status

A widespread but unobtrusive resident with some passage in autumn. This species is generally not as well recorded as it should be, though it is certainly resident throughout the city of Cambridge. Easy (pers. comm.) states that this species used to gather in small flocks in fenland after the breeding season but does so far less often now. Other records of small flocks are occasionally received.

Breeding Status

It is shown in the *Breeding Atlas* (Sharrock) to be widespread but there is little or no information on density. Personal experience, together with that of Easy (pers. comm.), suggests that most villages have a few resident pairs but there can be no doubt that this species is under-recorded. At Coton (Bucknell 1977) Hayley (Conder 1975) and in the Botanic Garden (Harper 1982) 1-2 pairs breed.

Ringing Results

A bird ringed at Little Eversden in November 1962 was found in the Gironde district of FRANCE in January 1963 showing that migration, reported by other counties (c.f. Birds of Kent), takes place through Cambridgeshire, although probably on a small scale.

Cetti's Warbler   *Cettia cettia*

Unrecorded prior to 1977.

Present Status

A most uncommon resident, found at only a handful of sites.
This species arrived in England in 1961 and was first recorded breeding (in Kent) in 1972 (Bonham and Robertson 1975). In Cambridgeshire it was first recorded in November/December 1977. There were no further records until 1980 when it was seen and heard at Wicken Fen. Subsequently it has been regularly recorded there and once or twice at one or two other wetland sites, including the Ouse Washes. Hard winters, however, seem to limit the population expansion and reduce the numbers rather drastically.

Breeding Status

At Wicken Fen 3-4 pairs bred regularly between 1980 and 1984 with 3-7 birds ringed annually including several juveniles. However, after the hard winter of 1984/85 only a single bird remained and that disappeared after 28 April so breeding has ceased (temporarily?) at that site. At Melbourn a single bird was present throughout the summer of 1985 but breeding was not proven.

Ringing Results

One of the first birds to arrive at Wicken Fen in 1980 had been ringed the previous year at Hoddesdon (Herts). A young bird from Wicken was found at Newbourn (Suffolk), the summer after it fledged. Both these birds show how the expansion has taken place.

BONHAM P.T. and ROBERTSON J.C.M. The spread of Cetti's Warbler in N.W. Europe. Brit. Birds 68. 1975.

Grasshopper Warbler   *Locustella naevia*

Pre 1934

Evans described this species as 'plentiful in Wicken Fen but an uncommon summer visitor elsewhere'. Lack more or less concurred saying that it was found breeding at Wicken and neighbouring fens but was very local.

1934-1969

Up until the 1950s there were only one or two records each year mainly from Chippenham, Fulbourn and Wicken Fens and Fowlmere watercress beds, and birds were noted as breeding at all these sites. In the 1950s and 1960s new sites were reported although some were probably in use earlier. These included Cambridge sewage farm, Waterbeach GP, Kennett, Shepreth L-Moor, Hauxton GP, Whittlesey, Milton GP, Ditton Park Wood, etc., but few if any of these were to become traditional sites, being suitable breeding areas only for a few years. In the late 1960s there were estimates of 20-30 pairs in the county including 12 singing males at Wicken Fen in 1965.

Present Status

A local and and surprisingly restricted summer resident.
Found at traditional sites around the county. Some passage birds are noted and some males take up territory at small sites unsuited to breeding.

Breeding Status

In 1971 a county total of 44 singing males at five or six sites was perhaps the best record in recent years. The present population varies between 20 and 40 pairs. The major sites are Wicken Fen (maximum 10 1980), Fowlmere (maximum 11 1982), Fulbourn Fen (up to 7 1980) and Chippenham (up to 4 1982). Many other sites are used irregularly with 1-2 pairs present, and while numbers in 1984 suggested a decline with a reported total of only 12 singing males, in 1985 a possible 29 singing males were discovered in the county.
Earliest date: 14 April (1952 Wicken Fen)
Latest date: 29 September (1973 Wicken Fen)

Savi's Warbler   *Locustella luscinioides*

Pre 1934

Evans stated that it bred in the county until the draining of the fens about

1849. Lack mentioned the fact that this species was first identified in Britain in Cambridgeshire in about 1840, but the fenmen had long recognised it as different from the Grasshopper Warbler and called it the 'brown, red or night' reeler. It disappeared in about 1850.

1934-Present status

A rare summer visitor.

It re-appeared first in 1954 and there have been seven records since, listed below:

1 One at Wicken Fen was heard and seen through the period 3 June-mid August 1954. (This was the first record for Great Britain since 1916.)

[One was heard and seen 28 April-13 May 1956 at Wicken Fen. Despite being seen by several observers this record was never considered more than a strong possibility.]

2 A single male at Wicken Fen, 7 July 1969.

3 A male singing on the Ouse Washes 19-26 June and 17 July 1978 was thought to be alone and therefore not breeding.

4 Present on the Ouse Washes from May to August 1979 with three singing birds on 27 July.

Savi's Warbler on Ouse Washes, 1979.

5 A male at Wicken Fen from 12 April to 26 July 1980. At times a second male was present. It was thought to have bred.
6 One at Wicken Fen from 28 April to 23 May 1983.
7 A single bird 'reeling' at Fowlmere watercress beds during May 1985.

Moustached Warbler   *Acrocephalus melanopogon*

An extremely rare vagrant.
The infamous and much discussed record of a pair which bred and raised young at the ballast pits at Cambridge sewage farm in 1946 (Hinde and Thom 1947) remains the only such record for Great Britain. Despite considerable scrutiny on many occasions (Sharrock 1981) of the evidence of identification this record remains totally substantiated.

HINDE R.A. and THOM A.S. The breeding of the Moustached Warbler in Cambridgeshire. *Brit. Birds* 40. 1947.
SHARROCK J.T.R. *Birds new to Britain and Ireland.* T.& A.D. Poyser. 1981.

Moustached Warbler nested in Milton in 1946.

Aquatic Warbler   *Acrocephalus paludicola*

An extremely rare vagrant.

Three records listed below:

1 A bird at March, 4-16 May 1953. The presence of a second bird was suspected.
2 One was seen at Peterborough sewage farm, 28 August 1954.
3 One at Cambridge sewage farm, 1 August 1955.

Sedge Warbler   *Acrocephalus schoenobaenus*

Pre 1934

Evans described this species as a 'very common summer resident'. Lack went further and stated that it was abundant and widely distributed. He considered that in the fens it was more abundant than in any other agricultural area in Britain.

1934-Present Status

A common summer resident.

Found throughout the county but more concentrated in the wetland areas, particularly in fenland. In general birds arrive in mid April and depart late September/early October. Easy (pers. comm.) considers that the south Cambridgeshire population increased during the 1960s and that later reductions (Fig.46) merely brought the population back to former levels.

Breeding Status

Widespread and in large numbers at such sites as the Ouse Washes, Wicken, Fulbourn and Chippenham Fens, Fowlmere watercress beds and various gravel and ballast pits. However, since 1970 there have been continuous reports of numbers decreasing as noted in 1974 on the Ouse Washes by Sorensen (1974) and while, after the first decline, there was a sign of stability, in recent years numbers have fallen again as shown by the number of breeding pairs at Fowlmere (1981: 110 singing males, 1982: 73, 1984 down to 32, 1985: 45). At Wicken Fen the numbers ringed (presented as a percentage of the total to allow for catching variation) have fallen since the early 1970s (see Fig.46) and the BTO Common Bird Census shows that population levels in 1985 were a quarter of those in 1968. In the fenland area, where this species has been traditionally strong, there has been a marked decline probably due in part to habitat loss.

Ringing Results

As with the Reed Warbler, considerable information has been collected. Foreign recoveries show a similar pattern (see Reed Warbler), however, one bird was found at Le Migron, FRANCE, only a week after being caught at Wicken Fen. Others were recovered in BELGIUM, PORTUGAL and MOROCCO. The English recoveries also show the migration routes in and out of the county with birds controlled at such places as Dungeness (Kent), Easton, (Hants), Penzance, (Cornwall), and one at Slapton (Devon) which had gained 5.3 grams in five days which represents about a 50% weight gain. A bird from the north-west of England (Carnforth, Lancs). ringed in July 1977, passed through Wicken Fen a month later raising an interesting possibilty that some birds cross the country to leave by the south-east coast and make the shortest sea crossing.

Figure 46. Reed Warbler and Sedge Warbler – the number of birds ringed at Wicken Fen expressed as a percentage of the overall catch 1971-85.

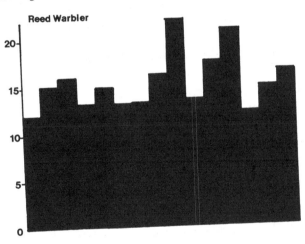

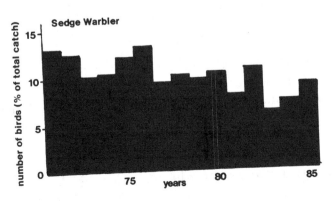

Earliest date: 29 March (1981 Fen Drayton GP and Wicken Fen)
Latest date: 28 October (1979 Ouse Washes)

SORENSEN Jeremy. Population trends of some small passerines on the Ouse Washes. *Camb. Bird Club Report* 48. 1974

Marsh Warbler   *Acrocephalus palustris*

Pre 1934

Evans reported a 'nest taken' in the county and Lack stated that there were a few summer records including breeding at Wicken Fen in June 1859 and at Whittlesford in June 1909. The first of these records is surprising since Fisher (1966) stated that this species was only first distinguished from the Reed Warbler in Britain in 1871!

1934-Present Status

A rare visitor in summer.

The records are presented by the years in which they were reported.

1942 An immature at Wicken Fen, 15 July.

1952 A pair bred at Hauxton GP where they were noted from 3 June.

1953 One singing at Hauxton GP and two adults in July. (The 1953 *Camb. Bird Club Report* noted many unconfirmed records between 1937 and 1953.)

1954 A pair nested at Hauxton GP and there were several other unconfirmed reports.

1956 A new colony was discovered with one nest and up to 10 birds found.

[1959 A nest was produced at a local meeting said to be of this species from Cottenham!]

1962 One was heard singing at Hauxton GP.

1974 One was heard singing at Ely beet factory, 17 May.

1976 One was trapped and ringed at Fowlmere, 31 May.

1982 A male singing throughout June on the Ouse Washes.

1983 One at Fowlmere on 7 June.

This species is summarised in this unconventional way because it is undoubtedly one of the most difficult to assess. The difficulties of positive identification, even in the hand (Williamson 1968), are legendary and without vocal evidence records are highly questionable. Easy, writing in the 1984 *Camb. Bird Club Report*, summarised the position admirably when he stated that some of these records are now suspect while others that have been rejected may now be more acceptable!

FISHER J. *The Shell Bird Book*. Edbury Press and Michael Joseph. 1966.
WILLIAMSON K. *Identification guide for ringers 1. Cettia, Locustella, Acrocephalus, and Hippolais*. BTO. 1968.

Reed Warbler   *Acrocephalus scirpaceus*

Pre 1934

Evans described this species as a 'local summer visitor, plentiful where it occurs' along river valleys and around ponds. Lack concurred with this description and added that the main strongholds were in the fen country and the southern parts of the county.

1934-Present Status

A common summer resident.
Found from late April to late September and often into October in most places where Phragmites beds of any size occur; the size of each colony depends on the area of the reeds. This species has been the subject of intensive study in the county, particularly at Wicken Fen (see Wicken Fen Group Reports) and at Fowlmere watercress beds.

Breeding Status

It is not as widespread as the Sedge Warbler due to its rather exact requirements, nevertheless the *Breeding Atlas* (Sharrock) shows it to be found in most 10 km squares (even though it is virtually absent from the upland areas of the county). Recent counts of singing males at Fowlmere show a population of between 50 and 150 pairs; probably the largest colony though is at Wicken Fen where between 261 and 774 have been ringed per annum since 1970 (see Fig.46).

Ringing Results.

Since this species has been the subject of an intensive ringing programme at both Wicken Fen and Fowlmere it is hardly surprising that there is a comparatively large number of ringing recoveries. These can be divided into two types. First the foreign recoveries showing the route to the wintering areas; these include one in FRANCE, two in SPAIN, one in PORTUGAL, two in MOROCCO and one in MAURITANIA. Secondly the English recoveries; these are all in places to the south of Cambridgeshire and show birds generally leaving and arriving via the south coast, particularly in the Sussex/Hampshire area. One or two birds were ringed one summer in Cambridgeshire and subsequently controlled elsewhere, two in the west country.

Earliest date: 14 April (1985 Wicken Fen)
Latest date: 27 October (1979 Wicken Fen)

Great Reed Warbler   *Acrocephalus arundinaceus*

An extremely rare vagrant
Three records listed below:

1 One caught and ringed at Wicken Fen, 21-22 May 1971.
2 One singing at Purl's Bridge on the Ouse Washes, 24 May-18 June 1981.
3 One singing on the Ouse Washes, 7-9 July 1982.

Barred Warbler   *Sylvia nisoria*

An extremely rare vagrant.
Three records listed below:

1 One shot by a porter at Newnham near Queens' College, Cambridge around 1839. [Lack]
2 One caught and ringed on Fleam Dyke, near Fulbourn, 11 October 1970.
3 A juvenile caught and ringed at Wicken Fen, 22 September 1979.

Icterine Warbler

An extremely rare vagrant.
Two records listed below:
[A bird at Milton GP on 19 August 1966 was considered almost certainly to be of this species but due to the short period of observation this record remained unacceptable.]

1 One at Wisbech, 6 August 1970.
2 A singing male on the Gog Magog Hills between 16 June and 16 July 1980 was the subject of considerable interest since it exhibited many features of the Melodious Warbler (Grant and Medhurst 1982).

GRANT P.J. and MEDHURST H.R. The Wandlebury Warbler. *Brit. Birds* 75. 1982.

Lesser Whitethroat   *Sylvia curruca*

Pre 1934

Evans described this species as 'a summer visitor, more plentiful with us than in most districts'. Lack stated that it was not uncommon in the more southerly parts of the county.

1934-Present Status

A moderately widespread summer resident.
This species, which is present from late April to late September, has a rather uneven distribution across the county. There was a period of time in the late 1970s and early 1980s when the population levels fell rather alarmingly (see Fig.47) but there are signs of a recovery recently. As with other migrants the drought conditions in north Africa seem to have depleted breeding stocks.

Breeding Status

The *Breeding Atlas* (Sharrock) shows some 10 km squares not occupied and distribution is less complete than for the Whitethroat. Being more a bird of scrubland and thickets its distribution is more suited to the south and central parts of the county. On the Coton census site this species was recorded in 1974 after seven blank years, although it bred annually thereafter with between one and four pairs (Bucknell 1977).

Figure 47. Lesser Whitethroat – the number of birds ringed at Wicken Fen 1971-85.

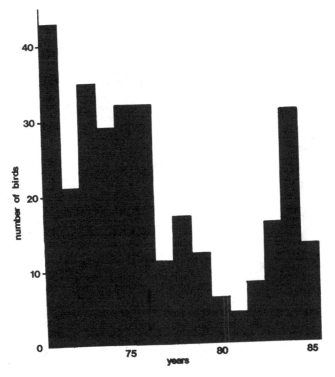

Ringing Results

Two birds ringed in Cambridgeshire in July were found in Northumberland and BELGIUM in May and April respectively of subsequent years. A first-year bird ringed in August 1963 in Berkshire was found at Great Wilbraham in June 1966, presumably breeding.
Earliest date: 14 April (1985 Wicken Fen)
Latest date: 8 November (1956 Milton GP)

Whitethroat   *Sylvia communis*

Pre 1934

Evans described this species as a 'common summer visitor'. Lack went further, stating that it was an abundant and widely distributed summer visitor.

1934-Present Status

A summer resident, but no longer the common and often abundant bird of arable land.
Without being hard to find this species, as a result of a dramatic crash in numbers due to drought in winter quarters (Winstanley, Spencer and

Figure 48. Whitethroat – the number of birds ringed at Wicken Fen 1971-85.

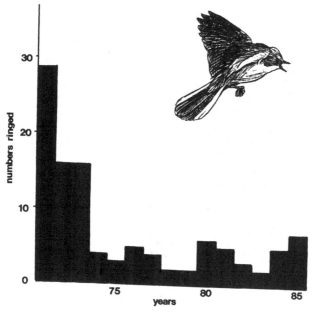

Williamson 1974), has vanished from many of its former haunts, and where it remains the numbers have been reduced rather more permanently than was originally anticipated (see Fig.48). This is particularly true of fenland.

Breeding Status

It is shown in the *Breeding Atlas* (Sharrock) to be widespread. Records that have been received, together with ringing returns, suggest that there has been no re-colonisation and that while distribution is rather uncertain, density is greatly reduced. Easy (pers. comm.) states that on his 50 acres in Milton there has been a deterioration from a population of 12 pairs in the 1960s down to two pairs. Bibby (1970), using Common Bird Census data from Soham and Ely suggested that this species was declining due to the loss of hedgerow habitat before the catastrophic fall. At Coton the number of pairs fell from 15 in 1967 and 1968 to a maximum of nine and a minimum of two in subsequent years (Bucknell 1977).

Ringing Results

An adult female ringed at Carlton in May 1966 was found at Lignieres, FRANCE, in May 1967. A bird ringed in September 1980 at Wicken Fen was found at Titchwell, (Norfolk), in June 1981.
Earliest date: 1 April (1967 Cambridge sewage farm)
Latest date: 28 October (1947 Oakington)

BIBBY C.J. The Common Bird Census. *Camb. Bird Club Report* 44. 1970.
WINSTANLEY Derek, SPENCER Robert and WILLIAMSON Kenneth. Where have all the Whitethroats gone? *Bird Study* 21. 1974.

Garden Warbler   *Sylvia borin*

Pre 1934

Evans described this species as a 'plentiful summer visitor'. Lack stated that it was a summer resident to the more wooded parts of the county especially in the west, and was not very numerous but well distributed considering the scarcity of woodland.

1934-Present Status

A summer resident.
Present from early May to early October. Far less numerous than the Blackcap and almost absent in the fens, particularly the open areas. Ringing catches suggest probable passage through the county in spring

and autumn, but on a small scale. Numbers ringed at Wicken Fen suggest a stable population.

Breeding Status

The *Breeding Atlas* (Sharrock) shows that the distribution is very much centred in the south of the county but that this species is not numerous. Conder (1975) stated that the population in Hayley Wood was 3-4 pairs.
Earliest date: 17 April (1949)
Latest date: 18 October (1970 Fleam Dyke)

Blackcap   *Sylvia atricapilla*

Pre 1934

Evans described this species as a 'plentiful summer visitor'. Lack stated that it was a summer resident, commoner and more widely distributed than the Garden Warbler. One was recorded by Evans in the winter of 1885.

1934-Present Status

A widely distributed common summer resident.

Over the last 30 years the status in general has changed very little except that there has been an increase in the number of reports of wintering birds in recent years with up to 15 in some years. At Stapleford wintering birds, sometimes a male and a female, have been noted around a bird-table for eight consecutive years (B. Harrup pers. comm). However, a survey by Leach (1981) showed Cambridgeshire to have very few records in comparison with other counties.

Breeding Status

The *Breeding Atlas* (Sharrock) shows the Blackcap to be a more generally distributed breeding species than the Garden Warbler and since it finds overgrown pits, rural situations and suburbia to its liking it is found even in small pockets of fenland. The population in Hayley Wood was estimated as up to 10 pairs (Conder 1975), in the Botanic Garden two pairs were recorded (Harper 1982) and on the Coton census site 1-3 pairs (Bucknell 1977).

Ringing Results

Three birds, ringed at Lode in 1976, at Wicken Fen in September 1979 and at Wilburton in May 1980 have all been found in MOROCCO. Two birds which departed from Beachy Head (Sussex) in September 1971 and

1972 were both at Wicken Fen in the summer of 1973. Evidence of spring passage is provided by a bird ringed at Dernford Fen, Sawston, in April 1979 that was in Lincolnshire 15 days later.

LEACH Iain H. Wintering Blackcap in Britain and Ireland. *Bird Study* 28. 1981.

Dartford Warbler   *Sylvia undata*

An extremely rare vagrant.
One record only:
A single bird was reported at Great Abington in 1870.

Wood Warbler   *Phylloscopus sibilatrix*

Pre 1934
Jenyns stated that it bred at Bottisham in the early nineteenth century. Evans reported that this species had been seen in Cambridge and Bottisham and that a specimen was in the Cambridge University Zoology Museum. Lack described it as a scarce summer visitor with one or two breeding records from the Cambridge area around 1910-15. Subsequently it was only seen on passage.

1934-1969
In 1942, the Camb. Bird Club woodland survey showed that this species had been under-recorded since five pairs were reported to be nesting (two at Arrington, two at Chippenham, and one at Hardwick) but apart from these the only other records were of passage birds, and those not annually. In 1956 two pairs bred at Babraham, and over the years there were records of birds singing for a few days in May on the 'Backs' and at Cherry Hinton but nesting was never established. After 1956 records were sparse and sporadic.

Present Status
A scarce passage migrant, particularly in spring, and a sporadic breeding species.
This species may well have been (and still be) under-recorded in the previous period since there have been annual records since 1973, with several birds noted at many sites all away from the open fenland.

Breeding Status
A few birds still nest irregularly but suitable sites are rare in Cam-

bridgeshire and males often hold territory but fail to attract a mate. The *Breeding Atlas* (Sharrock) shows a westerly national breeding distribution with a marked absence in East Anglia and Cambridgeshire in particular. Breeding in recent years has only been recorded at Hayley Wood and other boulder clay woods.

Earliest date: 7 April (1969 Ouse Washes)

Latest date: 15 September (1980 Ouse Washes)

Chiffchaff   *Phylloscopus collybita*

Pre 1934

Evans described this species as a 'widely distributed summer visitor, nowhere very abundant'. Lack stated that it was a fairly common summer visitor breeding in most woodlands of the county. One was shot on 2 December in 1856 at Batesbite lock.

1934-Present Status

A fairly common summer resident more numerous on passage. Present from late March to early October. Fairly widely distributed, but being a bird of the canopy it requires tall mature trees and is thus somewhat restricted in the areas of open fenland. Wintering has been recorded, increasingly over the last 15 years, but not as frequently as the Blackcap. Numbers ringed per annum at Wicken Fen since 1971 have remained remarkably stable (the more so when fluctuations in the total catching effort are taken into account).

Breeding Status

Widespread but rather thinly distributed mainly in the southern part of the county. It thrives in the suburbs of Cambridge and in the surrounding districts and is one of the few species adapted to conifer plantations. It is very restricted in fenland. Conder (1975) estimated 5-10 pairs in Hayley Wood, there was only a single record in the Botanic Garden in 1982 (Harper 1982) and this species was not recorded on the Coton census site.

Willow Warbler   *Phylloscopus trochilus*

Pre 1934

It was described by Evans as a 'moderately common summer visitor'. Lack stated that it was an extremely common and widely distributed summer resident. He added that it was more common in the river valleys in the south than in the fens and on the chalklands.

1934-Present Status

A very common and widespread summer resident.
Present from early April to September with some passage birds in spring and autumn. The Willow Warbler is the most common of the *Phylloscopus* warblers but it is absent from many arable areas of the fenland. It is, however, very abundant in the areas where there is any scrubland and at Wicken Fen it has been the third-most (recently with the decline of the Sedge Warbler the second-most) ringed warbler. In suburban areas this species is more thinly distributed.

Breeding Status

A widespread and abundant breeding species although it is poorly represented in the north of the county being absent from quite extensive areas of arable fenland. In Hayley Wood up to 30 pairs were thought to breed (Conder 1975), in the Botanic Garden the estimate was 1-2 pairs (Harper 1982) and between 2 and 9 pairs were counted on the Coton census site (Bucknell 1977).

Ringing Results

One ringed in Cambridge in June 1954 was found near Hythe (Kent) on 13 August of the same year, presumably about to depart on autumn migration. A bird ringed at Wicken Fen in August 1983 was controlled in Lincolnshire in August 1985 while another ringed in Lincolnshire in August 1984 was controlled at Wicken Fen in September 1985, indicating some autumn passage through the county.
Earliest date: 12 March (1957 Babraham)
Latest date: 10 October (1949 Girton)

Goldcrest   *Regulus regulus*

Pre 1934

Evans described this species as 'not very common'. Lack stated that it was 'scarce as are fir trees' and that it was found, and breds, where fir trees were present and also in fenland in winter.

1934-Present Status

A fairly common resident in all areas apart from fenland.
In the early 1940s this species was relatively unusual but was showing signs of increasing and was to be found in most of the larger woodlands, e.g Madingley Wood and in Cambridge in places like the Botanic Garden.

This species has shown a steady increase since the 1960s, mainly due to the maturing of ornamental conifers that have become so popular in gardens, leading to common occurrence in suburbia and in the villages around the county. Numbers in winter have also increased as has the area of distribution. Thorne and Bennett (1982) stated that this species was found mainly at Wicken Fen in late autumn and winter and ringing returns show a considerable increase in numbers in recent years, almost regardless of the total catch.

Breeding Status

The *Breeding Atlas* (Sharrock) shows it to be present only in the south of the county; it is certainly more widespread now as a result of the developments mentioned above.

Firecrest   *Regulus ignicapillus*

Pre 1934

Evans stated that the original record for Great Britain was of a bird taken near Cambridge in 1852 but that this was considered doubtful. Lack added that the bird in question was not identified correctly.

1934-1969

Batten (1973) reported that in Hampshire breeding began in 1961 and that the spread northward followed over the next ten years. The records for Cambridgeshire fit that pattern in general.

Six records in this period are listed below:

1 One at Mepal, 10 November 1956.

Figure 49. Firecrest – monthly distribution of all records.

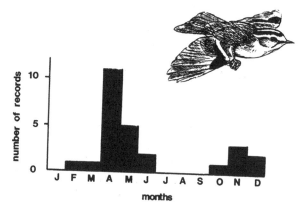

2 One found dead at Bassingbourn, end of November 1959.
3 One at Milton GP, 1-3 April 1964.
4 One at Dimmock's Cote, 11 April 1964.
5 One at Girton, 3 April 1969.
6 One at Coton, 30 April 1969.

Present Status

An increasing passage visitor, with breeding reported in 1982. Since 1970 there have been a further 19 records but they are not listed since it is quite clear that this species is now a regular, out of breeding season, visitor with annual records since 1980. There is a slight bias towards the spring (April in particular) and the monthly distribution of the records is shown in Figure 49. By far the most sensational of all the records is from the Ouse Washes on 3 June 1984 when one was discovered in the same bush as the county's first Thrush Nightingale!

Breeding Status

In 1982 a pair was presumed to have bred at Newnham College.

BATTEN L.A. The colonisation of England by the Firecrest. *Brit. Birds* 66. 1973.

Spotted Flycatcher   *Muscicapa striata*

Pre 1934

Evans described this species as an abundant summer visitor. Lack, on the other hand, stated that it was not abundant but well distributed and that it had declined around 1912.

1934-1969

During this period there was little change in status with the main stronghold being around the suburban areas, the villages and the larger woodlands.

Present Status

A reasonably common summer resident.
Present from mid May to late September. In the 1970s there were reports of a decline in numbers (see Wicken Fen ringing numbers Figure 50) but the 1984 records suggest that there is now a recovery at that site. Nationally this species has been noted to be in decline over this period with Common Bird Census results showing the population falling.

Breeding Status

This species has a more widespread distribution than many insectivorous species according to the *Breeding Atlas* (Sharrock). As suggested above, suitable sites are provided by the effects of human habitation and this species is often found in open-fronted nestboxes. Up to five pairs breed in Hayley Wood (Conder 1975), yet a maximum of ten pairs breed in the Botanic Garden (Harper 1982). A single pair occasionally bred on the Coton farmland site.

Ringing Results

Two birds ringed in the county have been found in MOROCCO, one rather curiously in the middle of July, but this may have been a 'long dead' bird.
Earliest date: 15 April (1967)
Latest date: 11 October (1978 Comberton)

Pied Flycatcher   *Ficedula hypoleuca*

Pre 1934

Evans stated that this species 'has occurred in May at Hinxton in 1836' and that the specimen was in the Saffron Walden Museum. Lack quoted two other records from 1832 and 1897.

Figure 50. Spotted Flycatcher – the number of birds ringed at Wicken Fen 1971-85.

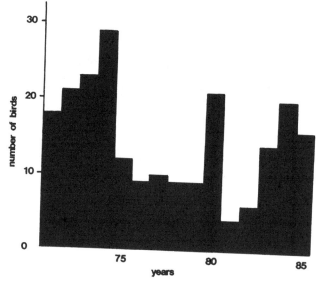

1934-1969

There were nearly 40 records in this period with a seasonal ratio of 2:1 autumn to spring. Most records are from the southern half of the county and involve only single birds, although exceptionally four were seen on the Ouse Washes in September 1961.

Present Status

An unusual and irregular passage migrant in spring and autumn. Around 40 records since 1970 make an overall total close on 80. Most are of single birds and apart from those at the Ouse Washes, Ely beet factory and Wicken Fen come from the Cambridge plus 15 miles radius catchment area. Figure 51 shows the seasonal bias towards autumn and many of these records coincide with 'falls' on the east coast. It is very noticeable that in years when there are a number of records in autumn they all occur in a wave within the same month, for example in 1973 there were five records four of which were in September and the other on 22 August, and in 1984 five records all in the period 12-26 August. An unusual example of this was three spring records in May 1985 between the 1st and the 12th.

Earliest date: 26 April (1947 Cambridge)
Latest date: 20 October (1978 Meldreth)

Figure 51. Pied Flycatcher – monthly distribution of all records.

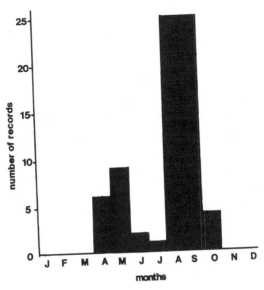

Bearded Tit   *Panurus biarmicus*

Pre 1934

Jenyns stated that it occurred in the fens and Burwell and Grantchester were mentioned. Evans described this species in the following terms 'no real evidence of anything'. Lack considered it to have been a resident formerly, but extinct by 1934. He mentioned that there were hints of breeding at or near Ely around the turn of the century and possibly at Whittlesey but no evidence was ever produced.

1934-1969

There were no records until 1957 (Vine 1960) when a male was present for a period at the end of the year at Cambridge sewage farm, marking the beginning of annual records. For the following seven years there were records in winter of 1-7 at one or two sites, usually Wicken and Fulbourn Fens. Easy (1967) considered these birds to be likely migrants from the Norfolk and Suffolk breeding colonies. In 1964 there was the first substantial count when up to 30 were present at Wicken Fen in November and December after which larger counts became the norm and by 1969 the maxima at the three favourite sites were: Wicken Fen, 40 (February 1966), Fulbourn Fen, 22 (October 1966), and Ely beet factory, 19 (1968).

Present Status

An unusual but regular winter visitor and highly local breeding species. In the early 1970s there was a further advance with an increase in the number of sites at which this species was found. In 1973 after what must have been an excellent breeding season in the stronghold of Norfolk and Suffolk, birds were found in autumn and winter at 12 wetland sites all around the county with up to 45 at Ely beet factory, and up to 30 at Wicken Fen. In 1974 oversummering began. Bibby (1973) summarised the status at this point and showed from ringing that the origin of these birds (and those in Huntingdonshire) was indeed the established breeding colonies in East Anglia and two birds were caught in successive winters showing a definite migratory pattern. The enlarged winter presence has continued with birds seen regularly at Wicken and Fulbourn Fens, Ely beet factory, Fowlmere, Ouse and Nene Washes, Fen Drayton and Waterbeach GPs. Numbers: up to 30 at Wicken Fen and Ely but elsewhere usually 1-10.

Breeding Status

Vine (1960), Easy (1967) and Bibby (1973) all predicted, with varying degrees of hope, the return of this bird as a breeding species. Surprisingly, many other counties were colonised before Cambridgeshire with proven records in Kent and Essex before 1966, in Lincolnshire, Hampshire, Dorset and Hertfordshire before 1970, and in Yorkshire, Sussex, Lancashire and Glamorgan before 1974 (O'Sullivan 1976). In 1974 it possibly nested with birds present at one site and juveniles seen in late July at another. However, after several years of suspected breeding the first established record was at Wicken Fen in 1980. In 1981 it bred at Fowlmere but although it summered at Wicken there was no evidence of nesting. In 1982 it bred at a fenland site but not at either Fowlmere or Wicken even though birds were present in summer. In 1983 breeding was reported from Wicken Fen and two other sites, and in 1984 and 1985 it bred at only one of the 1983 sites. It is clear, therefore, that this species remains very sensitive and until regular breeding at more than one site or even at one site occurs its foothold is precarious.

Ringing Results

Bibby's efforts, combined with those of the Wicken Fen Group, have given great insight into the origin and movements of this species. Birds wintering at Wicken have been found at Titchwell, Norfolk, (the following winter), at Murston (Kent) (the following summer), and two at Pommerouel in BELGIUM the following March. Birds have come to the county from Murston (ringed in summer), Walberswick, Suffolk (ringed in summer), and Stodmarsh (Kent) (ringed in early September). A single bird ringed at Wicken Fen in July was found at King's Lynn, Norfolk, the following January.

BIBBY C.J. Bearded Tits in Cambridgeshire and Huntingdonshire. *Camb. Bird Club Report* 47. 1973.

EASY G.M.S. Bearded Tit in Cambridgeshire. *Nature in Cambridgeshire.* 10. 1967.

O'SULLIVAN J.M. Bearded Tits in Britain and Ireland 1966-1974. *Brit. Birds* 69. 1976.

VINE A.E. Bearded Tits at Wicken. *Nature in Cambridgeshire.* 1960.

Long-Tailed Tit   *Aegithalos caudatus*

Pre 1934

Evans in his 1904 summaries considered this species to be more of a visitor than a resident, a comment he later reconsidered. Lack, however, stated that it was the second commonest tit in Cambridgeshire being more numerous than the Great Tit. He added that it was rare on the fenland.

1934-Present Status

A common resident with a widespread distribution, it is even found in fenland except its most open areas.

Most records are of small flocks made up from several family parties numbering up to 30 although exceptionally 60-70 were seen together at Cambridge sewage farm in January 1945. In some winters (in some gardens) it can be attracted to bird-tables in small numbers and in the city of Cambridge small parties are noted particularly in autumn and winter.

Breeding Status

Widespread and common. The *Breeding Atlas* (Sharrock) shows a reasonably widespread distribution with only one or two squares vacant. Up to five pairs breed in Hayley Wood (Conder 1975), up to three in the Botanic Garden (Harper 1982) and 1-2 pairs on the Coton census site (Bucknell 1977).

Ringing Results

Most of the recoveries are of birds found within a few miles of their ringing locations. Two birds show greater movement, from Wimblington in November 1971 to Downham Market, Norfolk, in January 1974, and from Fowlmere in October 1973 to Northwood, Middlesex, in July 1974.

Marsh Tit   *Parus palustris*

Pre 1934

Jenyns described this species as abundant, but conversely Evans considered it local and by no means common. Lack rather as a final arbiter stated that it was regular in Cambridge but very sparsely distributed. It must be remembered when considering these comments that the Willow Tit was not recognised as a separate British species until 1897.

1934-1969

A difficult species to assess due partly to the identification problems and partly to the lack of information that persists with all passerines that are neither common nor uncommon. In the early part of this period breeding was reported from Chippenham Fen, Madingley Wood, Longstowe, Hayley Wood, Wimpole Hall and Wicken Fen, (although at this last site Thorne and Bennett considered it unlikely and suggest confusion with the

Willow Tit) and it was said to be common in the southern part of the
county up to the edge of fenland. In 1963 an estimate of 100 pairs was
considered likely as a county total.

Present Status

A sparsely distributed resident.
Found particularly in the southern half of the county mainly in the
woodland area centred on the boulder clay and unusual in fenland where
it only seems to venture in winter. Favoured sites include Hayley, Hardwick and Fordham Woods, Wandlebury, Stetchworth and Woodditton.

Breeding Status

The *Breeding Atlas* (Sharrock) shows a clear southern distribution. Probably many areas contain 1-2 pairs and certainly it is commonly found in
summer at the above-mentioned sites.

Willow Tit   *Parus montanus*

Pre 1934

Lack stated that this species was found at Cambridge, Madingley and on
the Gog Magog Hills and he suggested that it was overlooked but nevertheless more unusual than the Marsh Tit.

1934-1969

Because of identification problems this species like the last has to be
reviewed with some circumspection. There are but a few records up to the
mid 1950s and these are from areas geographically separated, such as Ely,
Madingley, Dernford and Chippenham Fens, representing a spread across
the whole of the southern part of the county. Mostly groups of 1-2 were
counted but some 'small flocks' were noted. In the 1950s the most prominent area from which records were received was Wicken Fen where 2-4
pairs were reported to be breeding, and the Chippenham Fen/Kennett
area. In the 1960s records from the south (Borley Wood) and the west
(Hayley Wood) were added to the above-mentioned sites, suggesting
overlap with the previous species.

Present Status

A scattered resident.
Found mainly in the areas of scrub/wetland, and in the ancient woods on
the boulder clay. Its stronghold is in the Wicken Fen, Chippenham Fen,
Ouse Washes area where it is not in conflict with the Marsh Tit. It is also

reported from the southern and western parts of the county at sites such as Hardwick and Hayley Woods, Linton and Sawston where there is some overlap with the previous species. In short it has a strangely discontinuous distribution.

Breeding Status

The *Breeding Atlas* (Sharrock) shows it to be of a more patchy distribution than the Marsh Tit but extending farther north. Up to 8 pairs were reported from Ditton Park Wood in the early 1970s and 4-5 pairs breed at Wicken Fen; elsewhere the numbers are probably in the region of 1-2 pairs per site.

Coal Tit   *Parus ater*

Pre 1934

Evans described this species as moderately common and Lack stated that it was not uncommon in Cambridge and the woods to the west but was scarce elsewhere.

1934-Present Status

A moderately common resident.
Found predominantly in the southern half of the county, it is absent from much of fenland and it is not common on the fenland fringes. This species is a bird of gardens and plantations and is more widespread than in the past for like the Goldcrest this species has benefited from the conifer planting of the post-war era, also being rather small, it is more often located by call.

Breeding Status

The *Breeding Atlas* (Sharrock) shows that the biggest gap in the whole of the UK and Ireland occurs in Cambridgeshire with distribution being restricted to the southernmost parts of the county and even there it is somewhat patchy. At the time of writing this may prove to be an example of an expanding species and it could well become more widespread. Up to three pairs breed in the Botanic Garden (Harper 1982).

Blue Tit   *Parus caeruleus*

Pre 1934

Evans described this species as 'common' and Lack stated that it was an abundant and widely distributed resident, commoner than in much of England.

1934-Present Status

A most abundant resident with a widespread population.

Common across the county, in woodland, scrub, wetland and suburbia, this species is even found in open fenland where it uses any tree/copse area to live, helped no doubt by the local human inhabitants! In 1949 there were the first records of attacking milk bottle tops (cardboard in those days), and one bird was said to have been responsible for tearing up the lecture notes of the Professor of Greek although how this came about history does not relate!

Breeding Status

Breeding is widespread and numerous. Conder (1975) estimated a population of 20-50 pairs in Hayley Wood. Harper (1982) stated that 11 pairs nested in the Botanic Garden, and on the Coton census site between 3 and 13 pairs were counted (Bucknell 1977).

Ringing Results

Ringing has shown this species to be highly sedentary for of thousands ringed in the county only two have moved outside its boundaries and they only to neighbouring Essex. Garden ringing shows that many individuals can visit a site over the course of a single day.

Great Tit   *Parus major*

Pre 1934

Evans described this species as 'common' and Lack stated that it was found in most parts of the county but that it was not as abundant as in other parts of England.

1934-Present Status

A common and widespread resident.

As with the previous species it is found all over the county but it is particularly numerous south of the fenland. In autumn and winter quite large gatherings occur at certain sites, particularly in the beechwoods of the chalk uplands where, feeding on the fallen beechmast, numbers of the order of 50 to 100 can be found in quite a small area.

Breeding Status

The Great Tit has almost as complete a breeding distribution as the Blue Tit but the *Breeding Atlas* (Sharrock) shows one or two unoccupied squares in the north of the county where the countryside is so open and predominantly arable farmland as to preclude all breeding possibility. Conder (1975) estimated between 20 and 50 pairs in Hayley Wood, Harper (1982) noted 10 pairs in the Botanic Garden and between four and nine pairs were counted on the Coton Census site (Bucknell 1977).

Ringing Results

These show, as with the Blue Tit, the essentially sedentary nature of the species with most of the distance recoveries being in neighbouring counties with two exceptions: a bird at Bicester (Bucks) in April 1975 which had been ringed at Fowlmere the previous August, and a bird from Freith (also Bucks) in December 1964 found at Chatteris in July 1967.

Nuthatch   *Sitta europea*

Pre 1934

Evans considered this species to be 'local and nowhere abundant' and Lack stated that it was not uncommon in Cambridge but rare elsewhere.

1934-1969

Reported from a few sites over this period, especially Chippenham Fen and Cambridge (particularly on the 'Backs' where breeding was recorded). Elsewhere irregular reports came from Abington (2-3 pairs bred in 1951), Harston, Milton and Bottisham.

Present Status

A resident with a scattered distribution.
Found throughout the year at some favoured sites in the southern half of the county, particularly Hayley Wood, Wandlebury, Stetchworth, Chippenham/Fordham and Cambridge. Recorded increasingly over the last 15 years with 13 sites mentioned in 1984, mainly in the east and west where woodland occurs. This species is absent from the fenland and even on the fringes it is rare with only a single record at Wicken Fen (Thorne and Bennett 1982) despite the close breeding sites.

Breeding Status

The *Breeding Atlas* (Sharrock) shows it to be extremely scarce in Cambridgeshire but in recent years records have been received from Chippenham (up to 10 pairs), Stetchworth (up to 2 pairs), Wimpole Hall, and Hayley and Fulbourn Woods.

Treecreeper   *Certhia familiaris*

Pre 1934

Evans described this species as 'fairly plentiful' and Lack considered that it might well be overlooked although he stated that it was rather local.

1934-Present Status

A moderately common resident.
It is confined to the southern half of the county and absent from the fens and recorded most regularly in the area within a 15 to 20 mile radius of Cambridge. It is also present in one or two fenland fringes such as Wicken and Littleport.

Breeding Status

The distribution in the *Breeding Atlas* (Sharrock) shows a large gap in the north and alongside the Lincolnshire border (which is in fact the largest such gap in England). The unsuitablity of the area is obvious. In the south it is widespread, breeding in the many wooded fringes to fields and in parts of suburbia. Up to five pairs were estimated to be breeding in Hayley Wood (Conder 1975).

Ringing Results

Ringing shows this species to be highly sedentary.

Golden Oriole   *Oriolus oriolus*

Pre 1934

Lack quoted four records:
One to be found in the Cambridge University Zoology Museum dating from between 1870 and 1871.
One shot near Meldreth in July 1872.
One seen at Wicken Fen, 30 April and 3 May 1899.
Finally, a male seen in Cambridge on 21 May 1933.

1934-1969

There were three records in this period when this species remained a very rare vagrant.

A first-summer male in Cambridge, at the end of May, 1964.
A first-summer male at Manea, 10 July 1966.
A male caught and ringed at Harlton, 9 May 1969.

Present Status

A rare vagrant with a single pair breeding since 1982.
There were three records in the 1970s:
A male from Wisbech sewage farm wandered into Cambridgeshire. 12 August 1971.
One was seen at Madingley, 19 May 1973.
A male was heard in full song at Wicken Fen, 27 June 1979.
In 1980 two males and one or more juveniles were seen at Wandlebury (Gog Magog Hills), on 16 July. This may have been a result of breeding in the county or merely a unique passage sighting. There were two, records in 1982, one on the Devil's Dyke in spring and one, maybe two at Purl's Bridge (Ouse Washes) on 24 July. In 1983 the first record was received of a pair breeding. The nest site has been used in every year since (1983-87). Apart from the birds involved at the breeding site there have been only three other records, all of single birds in May.
Earliest date: 30 April (1899 Wicken Fen)
Latest date: 24 July (1982 Purl's Bridge, Ouse Washes)

Red-Backed Shrike   *Lanius collurio*

Pre 1934

Evans described this species as a 'plentiful summer visitor'. Lack stated that it was a regular but rather scarce summer resident around Cambridge, Newmarket, Fleam and Devil's Dyke, the Gog Magog Hills and occasionally the fens, Wicken in particular.

1934-1969

In the immediate post-Lack period the status remained unchanged and into the 1950s the number of breeding pairs was probably within the range 10-20, mostly in the Devil's Dyke-Newmarket area. In 1955 the *Camb. Bird Club Report* noted a decrease and this marked the beginning of a period in which the number of pairs was more within the range 5-10. By 1960, however, less than five pairs were reported so that the main decline had taken a mere decade. Two or three pairs remained until 1966 when a single site was occupied. In general records diminished in line with the reduced population.

Present Status

An unusual passage visitor. The last breeding record was in 1980. Apart from nesting in 1970, 1971 and 1980 at the old site mentioned

Red-backed Shrikes nested at Coldham's Common in Cambridgeshire in the 1960s.

above there have been only seven other records, all of single birds and more often in July-August than in May. Previous sporadic breeding is unlikely to be repeated, particularly since the decline of this species is a national phenomenon, not merely a local effect, which Bibby (1973) considered most likely due to a combination of climatic changes and habitat loss.

Earliest date: 6 May (1951)
Latest date: 10 September (1951)

BIBBY C.J. The Red-backed Shrike - A vanishing British species. *Bird Study* 20. 1973.

Woodchat Shrike   *Lanius senator*

An extremely rare vagrant.
One record only:

A single bird was 'obtained' near Swaffham Prior before 1840. The specimen is in the Saffron Walden Museum. [Lack]

Great Grey Shrike   *Lanius excubitor*

Pre 1934

Evans described this species as a 'rare winter visitor' and Lack quoted about 25 records between 1837 and 1892.

1934-1969

Up until 1958 there were only seven records, all of single birds and almost all of short stay so that in this part of the period its status was much as described by Lack. From 1958 this species was recorded annually and by the early 1960s there were records of individuals staying for two or three months duration at any one site. By 1969 birds were seen at several sites each winter. All this shows the usual pattern of an increasing species.

Present Status

A regular winter visitor to several traditional sites.
Some birds spend long periods (November-March) at individual sites; those most favoured are Wicken Fen, the Ouse Washes, Fulbourn Fen, Fowlmere, Chippenham Fen and Ely beet factory. One or two less favoured sites are visited fleetingly, particularly during passage periods.
Numbers: almost invariably single birds, two were reported from Wicken Fen in the winters of 1979/80 and 1981/82.

Earliest date: 10 October (1973 Ouse Washes)
Latest date: 5 May (1979 Wicken Fen)

Jay *Garrulus glandarius*

Pre 1934

Evans described this species as 'not uncommon'. Lack considered it to be scarce and irregularly distributed. He added that it was found chiefly in wooded districts and had probably increased during and after the First World War.

1934-Present Status

A moderately common resident.

The Jay has a rather patchy distribution based mainly in wooded areas. In fenland it is relatively unusual although on the fringes one or two pairs remain in the few small wooded pockets. In the early part of the post-Lack period this species seems to have been quite scarce but into the 1950s there was a gradual increase. Simms (1971) showed that Cambridgeshire, apart from Rutland, is the least wooded county in England and it is therefore to be expected that this species would not be widespread. Furthermore, it has not made the transition to suburban habitat that other woodland birds have managed and thus remains isolated. It is, however, more widespread in winter when it can be seen in all areas up to, and including, the fenland fringes. In some years (e.g. 1983) there are 'invasions' from the Continent when larger numbers are reported in small parties of 1 to 5.

Breeding Status

The *Breeding Atlas* (Sharrock) shows it to be confined to the southern, eastern and western perimeters of the county where woodland is found and in parts of the central-southern district.

Ringing Results

A single bird, ringed at Wicken in August 1982, was found at Earith (Hunts) the following April.

Magpie *Pica pica*

Pre 1934

Evans stated that 'it has become rare' and Lack described it as irregularly distributed and numerous in the fens! Farren told Lack that it was common up to the turn of the century then declined before recovering during the First World War.

1934-1969

This species seems to be given to considerable fluctuation in Cambridgeshire. By the late 1940s records continually mentioned that it was increasing, in 1947 several pairs were said to nest on the Ouse Washes and, as Lack suggested, fenland records were as common as those from other areas. In 1952 a record of 55 in a field in Cottenham remains rather unique and the nearest comparable counts were 15-20; generally 1-2 were noted. The first sign of a reduction in numbers came in the early 1960s and this may have been part of the organo-chlorine disaster of that time. By 1964 it was said to be restricted to the orchards and areas of rough grassland and this situation prevailed through the remainder of the period.

Present Status

An increasingly common resident.

At present Magpies are more common than for some time but with a rather patchy distribution. Throughout the 1970s there were records from a few traditional sites but by 1980 numbers were beginning to increase and from being seen in 17-22 parishes in the 1970s Magpies are now found annually in over 40 parishes.

Breeding Status

Visitors to the county never fail to be astonished by the dearth of Magpies and a glance at the *Breeding Atlas* (Sharrock) shows that Cambridgeshire is about the only substantial part of England and Wales in which there are gaps in the breeding distribution. In 1973 it was recorded breeding in only 10 parishes, in 1975 15 nests were found but by 1983 there were ten pairs in the parish of Milton alone. Among 1986 records was that of a pair that bred in Hills Road, Cambridge, showing that suburbia is also providing nesting opportunities.

Nutcracker   *Nucifraga caryocatactes*

A very rare vagrant

Seven records, listed below:

1   An example of the thin-billed race (*N.c.macrorhynchus*) killed near Wisbech on 8 November 1859. [Lack]
2   One in Cambridge, 1 October 1928.
3   One at Whittlesey, 9-10 September 1968.
4   One at March, 6-11 September 1968.
5   One at Wisbech St Mary on 16 November 1968.

(Records 3 to 5 were all part of a remarkable invasion of thin-billed birds in the autumn of 1968, Hollyer 1970.)

6 One at Wandlebury, 24-26 October 1985.
7 One at Waresley, 24 November 1985.

HOLLYER J.N. The invasion of Nutcrackers in the autumn of 1968. *Brit. Birds* 63. 1970.

Jackdaw   *Corvus monedula*

Pre 1934
Evans described this species as 'common' and Lack stated that it was a common and widely distributed resident though not as abundant as the Rook.

1934-Present Status
A common and widespread resident.
Unlike the Jay and Magpie this species has made the transition to urban life most successfully. It is common both on its own and in the company of Rooks particularly in feeding flocks and at communal roost sites: Land-wade (3000), Harston (4500), Quy (3500) and Knapwell (5000). In the fenland area quite large numbers gather when farm activities such as ploughing and drilling provide feeding opportunities.

Breeding Status
The *Breeding Atlas* (Sharrock) shows a thoroughly widespread distribution even in fenland where deserted and farmland buildings provide additional nest sites to the traditional treeholes and crevices. In Hayley Wood Conder (1975) estimated 20 pairs.

Ringing Results
A bird ringed in Zealand, DENMARK, in June 1953 was found near Ely in March 1954.

Rook   *Corvus frugilegus*

Pre 1934
Evans described this species as 'abundant' and Lack stated that it was very common although more scarce on the chalk uplands and remote parts of the fens. In May 1934 490 nests were counted in Cambridge and in 1931 350 nests in Madingley Hall Woods. Winter roosts of up to 15 000 were counted at Madingley with others at Hayley Wood and at Dullingham.

1934-Present Status

A common and widespread resident.
Found throughout the county with some considerable roosts noted at certain traditional sites such as Landwade (up to 4500), Harston (up to 7500), Knapwell (6000) and at Quy Hall (4000-6000).

Breeding Status

1934-1969

In 1942 nests counted around Cambridge totalled 150 with 123 in elm, 14 in beech, 7 in lime, and 6 in London plane. The nesting habits of this species render it easy prey to the counters as well as the shooters and thus information has been collected for many years. In 1951 a census of south Cambridgeshire, organised by P.D. Sell, suggested a total of 99 rookeries in 60 villages with a total of 5970 nests. This showed an increase over the count for the national survey of 1944-45. In general the species increased up until the mid 1950s but the first signs of a decline became evident in the 1960s.

1970-1985

The national Rook census (Sage and Vernon 1978) showed the species to be declining, probably as a result of the combined effects of toxic chemicals, Dutch Elm disease and a loss of preferred grassland habitat. Of all the counties Cambridgeshire suffered the greatest decline with the population down 68% from the 1945-46 level and more dramatically actually 49% down since 1960-65 showing that the greater part of the loss was sustained within a ten-year period. Easy (1983) showed that in the fenland breeding colonies there has been a gradual but persistent decline since about 1953.

Ringing Results

A bird ringed as a nestling at Giethorn, NETHERLANDS, in May 1949 was found at Fowlmere in February 1951. A nestling ringed in Lithuania, USSR, in May 1983 was found dead at Lode in May 1985.

EASY G.M.S. The demise of the fenland Rook. *Camb. Bird Club Report* 57. 1983.
SAGE B.S. and VERNON J.D. The 1975 national survey of rookeries *Bird Study* 25. 1978.

Carrion Crow   *Corvus corone*

Pre 1934

Evans considered this species to be 'not uncommon' and the Hooded Crow (*C.c.cornix*) was said to be plentiful in winter. Lack stated that it was common and widely distributed particularly in fenland districts. Usually one or two were seen but occasionally there were small flocks in winter.

1934-Present Status

A common and widely distributed resident.
Found all around the county, usually in pairs or small parties (4-6) feeding in open country or around gravel pits and rubbish tips. Roosts of quite large numbers are reported at traditional sites with up to 280 at Coveney/Wentworth, near the Ouse Washes. In Cambridge it has become most conspicuous since the demise of the Rook. Hooded Crows are to be found in small numbers, and only in winter usually in northern (fenland) districts.

Breeding Status

The *Breeding Atlas* (Sharrock) shows a good, if not entire, 10 km square distribution in the county and there is no reason to suppose that this species has declined. On the Coton census site 1-6 pairs bred (Bucknell 1977) and on the Ouse Washes 25 nests were located in 1985.

Raven   *Corvus corax*

Unrecorded this century.
Jenyns noted this species as occasionally seen at or near Bottisham but becoming more scarce. A pair was reported to have bred between Teversham and Fen Ditton in October 1828 but it vanished as a breeding species shortly after and there have been no other records.

Starling   *Sturnus vulgaris*

Pre 1934

Evans described this species as 'common' and Lack quoted records of two huge roosts at Elsworth (150 000) and at Haverhill, Suffolk (100 000)

1934-Present Status

An abundant and well-distributed resident, a passage migrant and winter visitor.
Roosts of huge numbers have occurred at traditional sites and were the subject of much investigation in the 1930s and 1940s. These sites can be used annually or just occasionally and it is not uncommon for roosts to move from one site to another in mid winter or even to a third site in

spring. The maxima at regular sites are Wimblington/Ouse Washes 300 000-400 000 (October/November 1975), Weston Colville 390 000 (1982), Fulbourn Fen 200 000-250 000 (October/November 1972), Wicken Fen 200 000 (November 1980), and at many other sites counts of 10 000-100 000 have been received.

Breeding Status

Breeding is recorded in every part of the county as can be seen in the *Breeding Atlas* (Sharrock) with birds making use of every possible nest site, natural or man-made. Between 4 and 13 pairs were counted on the Coton farmland census site (Bucknell 1977), and up to 20 pairs in the Botanic Garden (Harper 1982).

Ringing Results

Ringing shows the source of the birds that visit the county and these are displayed in Figure 52. All seem to come from an ENE direction and were found in Cambridgeshire out of the breeding season.

Rose Coloured Starling   *Sturnus roseus*

A very rare vagrant.

Eight records, all listed below:

1 One shot at Royston, August 1830. [Lack]
2 One at Fulbourn, July 1856. [Lack]
3 One at Wisbech, August 1856. [Lack]
4 One referred to by Smoothy. [Lack]

Figure 52. Starling – foreign recoveries of birds ringed in Cambridgeshire.

5 One undated in Wisbech Museum from Outwell. [Lack]
6 One in Saffron Walden Museum labelled 'Wallington Cambs'. [Lack]
   (Wallington is just over the county boundary.)
7 One shot in an orchard in Melbourn, 19 June 1937.
8 An adult at Pemford Farm, Stapleford, 22 October 1944.

House Sparrow   *Passer domesticus*

Pre 1934
Evans considered it 'very common' and Lack described it as abundant and widespread.

1934-Present Status
An abundant and widespread resident.
Suffice to say that this ubiquitous species is every bit as successful in the fenland as elsewhere and forms large flocks, often associated with Tree Sparrows, although it rarely moves far from human habitation. Some roosts of up to 3000 have been noted around the county.

Breeding Status
Breeding is widespread.

Ringing Results
Ringing has been very restricted (it is not encouraged by the BTO) but shows the species to be generally sedentary. A bird ringed at Carlton in January 1968 was found in King's Lynn (Norfolk) the following May.

Tree Sparrow   *Passer montanus*

Pre 1934
Evans described this species as abundant and Lack considered it unusually abundant in the county north of Cambridge but occasional elsewhere.

1934-Present Status
A moderately common resident.
Unlike many other passerines the Tree Sparrow is as common in the open spaces of fenland as it is in any other agricultural part of the county. It is probably under-recorded by observers who do not bother to separate it from the ubiquitous House Sparrow. Quite large flocks (hundreds) form on arable land mainly out of the breeding season and roosts of similar size

form at suitable sites (e.g. Wicken Fen). Generally this species is not as common in the area south of Cambridge. It is not a bird of urban or suburban habitat and is therefore restricted to open country.

Breeding Status

As a breeding species it is well distributed, but it is nowhere particularly numerous. In woodland or scrub where nestboxes are provided this species will readily use them in good numbers. In Hayley Wood Conder (1975) estimated 10-15 pairs while on the Coton census site there was only a single sporadic pair (Bucknell 1977).

Ringing Results

Quite large numbers have been ringed due to both catching at roost and the readiness with which this species uses nestboxes (allowing nestlings to be ringed). Ringed birds show very little evidence of long-range movement. A bird from Dungeness (Kent), ringed in December 1960, was found at Wicken Fen in June 1962, and a bird ringed in Cambridge in February 1964 was found on the Newarp Light Vessel the following April. Curiously the Tree Sparrow does not have a very high retrap rate and this therefore suggests that there is considerable local movement in populations.

Chaffinch   *Fringilla coelebs*

Pre 1934

Evans described this species as 'very common' and Lack stated that it bred throughout the county with winter flocks of up to 50 birds.

1934-Present Status

A very common resident, passage migrant and winter visitor.
Found throughout the county, but it is more common in the areas south of the fens than in fenland itself. Large flocks (several hundred) gather in autumn and winter in or near the few beechwoods in the county (Wandlebury, Wort's Causeway, etc.) particularly in early autumn when there appears to be an influx of birds from the Continent associated with Brambling. Three dead birds sent to Dr J.M. Harrison in 1947 were considered by him to be of three different races, one English (*F.c.gengleri*), one Swedish (*F.c.coelebs*) and the last Continental (*F.c.hortensis*), and there are many records of observed passage in October/November and April. These observations, together with a single ringing recovery in

FINLAND, suggest the source of these visitors. This species has mastered the transition to urban and suburban habitat and can be seen and heard commonly throughout the year in most gardens.

Breeding Status

Breeding distribution in the *Breeding Atlas* (Sharrock) shows no unoccupied 10 km squares, although density is likely to be higher in the more favourable southern districts. Conder (1975) estimated a population of 10 pairs in Hayley Wood and Harper (1982) suggested there were between 11 and 16 in the Botanic Garden. On the Coton farmland census site Bucknell (1977) recorded between 6 and 11 pairs.

Brambling   *Fringilla montifringilla*

Pre 1934

Evans described this species as 'not uncommon in winter in many places, large flocks are often met with'. Lack stated the same although he implied that birds were restricted to the chalk/beech areas.

1934-1969

In this middle period the status changed very little. Most of the records before 1950 were of small numbers as part of mixed finch flocks and were close to Cambridge. From 1950 on there were many records of flocks up to 800 strong usually in the south-east and eastern parts of the county associated with beechwoods.

Present Status

A regular if somewhat locally distributed winter visitor. Present from October to April. Large flocks (several hundred) occur where food is concentrated (Wandlebury, Abington, Hildersham, Newmarket, etc.) and this species is not uncommon in the fens where good numbers (20-420) can gather at suitable sites such as Ely beet factory, Wicken Fen and the Ouse Washes; at the latter site this species together with the Chaffinch works the tideline picking up seeds deposited on the edge of the banks.

Ringing Results

A bird ringed at Six Mile Bottom in January 1960 was recovered at Yevre (Marne), FRANCE, in January 1963. One ringed at Wicken Fen on 10 April 1971 was found on HELIGOLAND 14 days later.

Earliest date: 23 September (1970)
Latest date: 8 May (1978 Ouse Washes)

Serin   *Serinus serinus*

An extremely rare vagrant.
One record only:

A single bird was seen in the Botanic Garden in Cambridge during May and June 1965.

Greenfinch   *Carduelis chloris*

Pre 1934

Evans considered this species to be common. Lack stated that it was common and well distributed with large flocks in winter especially along hedgerows.

1934-Present Status

A widely distributed and abundant resident.
Found throughout the county in large numbers, and as common in fen-land as elsewhere with birds feeding on the fringes of fields usually in small family parties. Also a most common urban and suburban species. Flocks of 200-400 form wherever food is abundant at all times of the year

Figure 53. Greenfinch – British recoveries of birds ringed in Cambridgeshire.

but particularly in winter, and many roosts, usually of 50-100 birds, form at suitable sites, often in hedgerows or scrubland fringes.

Breeding Status

A very common and extremely widespread breeding species. At least 15 pairs breed in the Botanic Garden (Harper 1982) and between two and six pairs bred on the Coton census site (Bucknell 1977).

Ringing Results

There is a very mobile population in the county but almost uniquely this movement is confined to the British mainland (one in Eire). Results suggest movement away from the area, mostly to the south and west, in the winter (see Fig.53).

Goldfinch   *Carduelis carduelis*

Pre 1934

Evans described this species as 'not uncommon' and Lack stated that it was widespread and increasing having been reported rare at the turn of the century.

1934-Present Status

A common and widespread resident.
Found throughout the county this species is more thinly distributed in summer. As elsewhere, large post-breeding flocks form in late August through September and early October feeding particularly on thistle seed, and these can occur almost anywhere and are commonly in the region of 100-200. In winter small parties of 5-20 are regularly encountered and this species is a common feeder in alders. Unlike the Greenfinch this species has not been able to adjust to the urban and suburban environment with great success although it is recorded on occasion. Generally, however, the open countryside remains its preferred habitat.

Breeding Status

Found in all the 10 km squares in the county (Sharrock). Unlike the Greenfinch, which seems to move away from suburban areas to breed, this species often finds this habitat more suitable in the summer and in Cambridge city is often to be seen in pairs feeding on areas of allotments. Two to four pairs bred on the Coton census site (Bucknell 1977) and 5-8 in the Botanic Garden (Harper 1982).

Ringing Results

Ringing shows this species to be more migratory than the Greenfinch with birds heading south for the winter. Recoveries from FRANCE and SPAIN, mostly within three or four months of ringing, echo the national pattern. Ringing dates suggest that these are first-year birds.

Siskin   *Carduelis spinus*

Pre 1934

Jenyns noted large flocks in 1825 and 1829. Evans described this species as an 'occasional winter visitor and sometimes in large flocks'. Lack restated the Evans distribution but added that there had been only three records since 1900.

1934-1969

Up until 1955 this species remained an irregular winter visitor with small numbers (1-10) recorded at one or two sites. In many years it was not reported at all. A single breeding record in 1951 was considered to have possibly involved escaped birds. The year 1955 saw the commencement of annual records whose number increased rapidly and dramatically so that by the end of the 1960s 1-20 were reported from several sites every winter (Bircham 1974). A pair at Wilbraham on 3 June 1962 may have been breeding in the area.

Present Status

An uncommon but regular winter visitor.

Found at traditional and other sites between October and April. The increase in records peaked around 1972 and has remained around that level since. The exact reason for this increase is probably a reflection of the increased national breeding population (Sharrock). Numbers are generally small 2-10 being most common, but at some sites 10-20 can be found. Numbers of 40-50 are recorded but must be considered somewhat exceptional. Favoured sites with their respective maxima are: Fowlmere watercress beds (up to 40) Coe Fen, Cambridge (up to 25), Shelford recreation ground (up to 20), Hildersham (up to 25) and Chippenham Fen (15-20). This species is rarely found in fenland but has been recorded at the Ouse Washes, Ely beet factory and Littleport, and at Wicken Fen. In 1982 a male was noted singing at Chippenham Fen on 21 May and at Wicken Fen a bird was seen on 8 July 1981 but no further evidence of breeding has been forthcoming.

Ringing Results.
Despite great efforts to catch this species in the early 1970s, and a total
ringed of over 100, the only information that has been gathered is that a
bird wintered two years in succession in Sawston, a bird ringed at Dernford Fen, Sawston, in February 1972 was found at Woodbridge (Suffolk), a month later, and a bird ringed at Epping (Essex), in March 1973,
was controlled at Fowlmere in March 1974.

BIRCHAM P.M.M. The status of the Siskin in Cambridgeshire. *Camb. Bird Club Report*
48. 1974.

Linnet   *Carduelis cannabina*

Pre 1934
This species was described by Evans as 'fairly common but somewhat
local'. Lack stated that it was the commonest finch in winter in the open
fen country where it occurred in large flocks and it was well distributed in
the breeding season.

1934-Present Status
A widespread and common resident.
Found throughout the county wherever agricultural land dominates.
Large flocks of up to 1000 form in areas where food is abundant, particularly in winter. This species has not, however, made any substantial
encroachment into suburbia and is generally only found in rural areas
where gardens back onto open land or on allotment sites where neglectful
holders unconsciously provide additional feeding opportunities.

Breeding Status
Widespread. Each 10 km square has a breeding record according to the
Breeding Atlas (Sharrock). Between five and eight pairs bred on the Coton
census site (Bucknell 1977) and in the Botanic Garden between 10 and 23
pairs have been recorded (Harper 1982). In 1985 33 pairs were counted
on the Ouse Washes and eight pairs at Fowlmere watercress beds.

Ringing Results
Movement away from the county in winter, shown in the recoveries,
indicates that some birds spend time in sunnier climes, FRANCE, SPAIN
and PORTUGAL proving popular!

Twite *Carduelis flavirostris*

Pre 1934

Jenyns noted large flocks in one cold winter period and Evans stated that it 'strays to the county in autumn sometimes in large flocks'. Lack described this species as a winter visitor with isolated records of 1-2 birds particularly since 1900.

1934-1969

In the 20 years from 1934 to 1955 there were only one or two records, and from 1955 to 1960 there was on average a single record per annum usually of between 1 and 10 seen in a part of the county north of Cambridge. However, in the 1960s the number of reports each year began to increase and by 1967, with more regular observation, this species was seen throughout the winter on the Ouse Washes and sporadically at other sites.

Present Status

An uncommon winter visitor from October to March.
This species is recorded regularly throughout the winter on the Ouse Washes but is seen only irregularly at other sites although there are several records each year. Generally numbers are in the range 1-10 but are usually higher on the Washes with a maximum of 300 between January and April in 1976. Found almost exclusively north of Cambridge, improvements in observation and identification, together with an expanding population, are possible reasons for the increased level of recording.
Earliest date: 15 August (1978 Ouse Washes)
Latest date: 4 May (1973 Milton)

Redpoll *Carduelis flammea*

Pre 1934

Jenyns considered this species to be 'not common', and Evans agreed describing it as 'scarce'. He added that a 'Mealy' specimen had been found at Hinxton in May 1836. Lack stated that it was not common except in Cambridge, where it bred, and in places such as Wicken and Chippenham Fens and at Fordham. He considered that it might be overlooked elsewhere, and noted an absence in the fens.

1934-1969

Records in this period suggest that it was neither common nor uncommon. A regular breeding population was noted in Cambridge and in 1955 was said to number five pairs, but it was not until the early 1960s that it was noted as increasing thus leading to records from a wider area of the county.

Present Status

A moderately common resident.

Found throughout the county in varying degrees. Some passage migrants and winter visitors. After the expansion of the population in the late 1960s (see for example Sorensen 1974) and early 1970s a stabilisation or even slight decrease has taken place. Numbers caught at Wicken Fen give a clear indication of the population trends of this species (see Fig.54). In winter, flocks of 200-300 can be found, however, parties 20-50 strong are more the norm. Birds can be seen in urban and suburban situations but nowhere very commonly. 'Mealy' Redpolls (*C.f.flammea*) are occasionally reported in winter.

Figure 54. Redpoll – the number of birds ringed at Wicken Fen 1969-85.

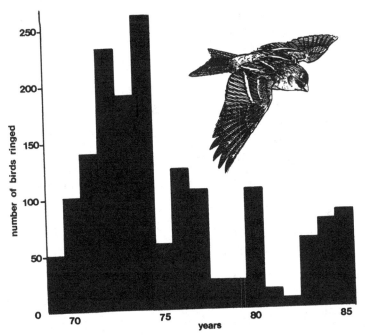

Breeding Status

The distribution according to the *Breeding Atlas* (Sharrock) is patchy although in areas where this species occurs several pairs are often present. Up to eight pairs breed in the Botanic Garden (Harper 1982), but on the Coton census site only a single occasional pair was recorded.

Ringing Results

The recoveries of ringed birds reflect the national pattern with a large number of foreign recoveries in BELGIUM and one or two in FRANCE. Most of these birds seem to have been caught in Cambridgeshire in autumn and found wintering on the Continent in the following year. There is evidence that some birds which summer in the county move to the southern parts of England for the winter.

SORENSEN J. Population trends of some small passerines on the Ouse Washes. *Camb. Bird Club Report* 48. 1974.

Crossbill   *Loxia curvirostra*

Pre 1934

Evans described this species as 'a rare straggler singly or in flocks' and Lack stated that it had occurred in recent years with about nine or ten records overall. The 1934 *Camb. Bird Club Report* contained an addendum to Lack which stated that the bird had bred in the county in 1926 which was the first such record.

1934-1969

Between 1934 and 1961 there were only four records, all of small numbers (1-8). An immigration in October 1961 led to many records over the following year, mostly of single birds, and subsequently a second breeding record near Newmarket in 1963. In 1964 a pair was seen near Newmarket Golf Course and near Cheveley and breeding may have occurred. Following this invasion period there was only a single record in 1966.

Present Status

An unusual and irregular visitor.

All the records in this period are listed below by year:

1972 An invasion led to records at: Botanic Garden, Cambridge (six in July), Whittlesford (seven on 6 July), Sawston (two on 26 July) and at Whittlesford again (one on 29 July).

1975 Two feeding on weed seeds near Coveney, 15 and 19 November.
1977 Two males, two females and eight juveniles, 9 June in Hayley Wood.
1981 One in Cambridge, 11 August.
1984 A juvenile at Toft, 5 May, and a pair on Newmarket Golf Course, 29 September.
1985 A single bird flew over calling at Croxton Park, 16 July.
1987 A male at Fowlmere, 14 July, and another on 13 September.

Pine Grosbeak   *Pinicola enucleator*

An extremely rare vagrant.
One record only:
A bird that was shot in Little Abington Vicarage garden, on 13 January 1882, is now in the Saffron Walden Museum. [Lack]

Bullfinch   *Pyrrhula pyrrhula*

Pre 1934
Evans considered this species to be 'fairly common'. Lack stated that it was not at all numerous being irregularly distributed and rare on the chalk uplands and the fenlands.

1934-Present Status
An abundant and widespread resident.
Commonly found throughout the county with a more thinly distributed population in the heart of the fenland. In parts of the county this bird is highly successful and, for example, around a small area of Sawston Hall ringing revealed a population of over 100 individuals each August/September period. While this species has not become common in urban and suburban districts it undoubtedly has learnt to make use of feeding opportunities that are inadvertently provided! Generally seen in pairs or small parties (2-5); large numbers are unusual and gather only when a concentrated food supply is available.

Breeding Status
The *Breeding Atlas* (Sharrock) shows a slight gap in distribution in the extreme north of the county, otherwise this species is widely found nesting and with a probably high density in suitable areas. Between one and five pairs bred on the Coton census site (Bucknell 1977) and up to five pairs breed in the Botanic Garden (Harper 1982).

Ringing Results

These show this species to be of a highly sedentary nature, the more so since large numbers are ringed and the only movement of any distance is a bird that was found at Histon in August 1964 which had been ringed at Woldingham (Surrey).

Hawfinch   *Coccothraustes coccothraustes*

Pre 1934

Described by Evans as 'rare and local' and by Lack as a not uncommon resident in Cambridge and the orchard districts. He added that it nested at Grantchester, Fulbourn and Pampisford. It was absent from the fens and probably overlooked.

1934-1969

It was regularly recorded at two or three sites in the early part of this period, notably Stetchworth/Newmarket, Abington/Hildersham, and Cambridge; and in the early 1950s there were breeding records from all of these areas. In 1950 it was seen in ten parishes and it was considered probable that it could be found all over south Cambridgeshire and in parts of the fens but just under-recorded. Following this period the reports began to decline and in most of the subsequent years there were a mere one or two. However, breeding continued to be recorded at Stetchworth.

Present Status

A local resident.

A shy and most secretive bird, this species is universally under-recorded (Sharrock) and Cambridgeshire is no exception. Casual sightings are reported at a rate of one to three per annum but from many different parishes as far apart as Chippenham and Queen Adelaide in the east and Hayley and Gransden Woods in the west and from Chatteris in the north to Sawston in the south. Numbers: usually 1-2 but exceptionally 6 at Haddenham in January 1981.

Breeding Status

This species is probably still breeding at its traditional sites although it has not been recorded for some time. The *Breeding Atlas* (Sharrock) shows the scattered and local nature of the records both nationally and in Cambridgeshire but since records have decreased despite a considerable increase in potental observers it has to be concluded that there has been a decline in what was always a meagre population.

Lapland Bunting   *Calcarius lapponicus*

A rare vagrant.

Sixteen records, all listed below:

1 One was sent to Leadenhall market from Cambridgeshire in February 1826. [Lack]
2 Onc was seen near Cambridge, November 1877. [Lack]
3/4 One was sent to Farren in the winter of 1892 and another soon after. [Lack]
5 One on the Ouse Washes near Oxlode, 3 January 1957.
6 Two, flying north across the Ouse Washes near Sutton, 23 December 1962.
7 One flying south on the Ouse Washes, 17 September 1968.
8 One on the Ouse Washes, 14 October 1969.

[A probable on the Ouse Washes, 19 December 1969.]

9 One on the Ouse Washes, 22 January 1978.
10 Ten plus on the Nene Washes, February 1978.
11 One flew over Purl's Bridge (Ouse Washes) calling, 16 December 1984.
12 A single flew over the Ouse Washes, 7 December 1985.
13 Small numbers (less than ten) wintered on, or around, the Nene Washes in 1986/87.

14 One on the Ouse Washes, 16 February 1987.
15 One or two were again present in the area of the Nene Washes from 18 October 1987 to the end of the year.
16 A male at Fen Drayton GP, 22 November 1987.

Snow Bunting   *Plectrophenax nivalis*

Pre 1934

Jenyns described a small flock on Newmarket Heath in 1826, one at Bottisham in February 1827, and one in Cambridge in January 1829. Evans considered it an 'uncommon winter visitor' and Lack quoted nine additonal records to those above making a grand total of twelve.

1934-1969

Between 1934 and 1950 this species was recorded only twice. Thereafter for 14 years it was seen annually with often two or three records each year. Numbers: generally 1-6 but in the November of 1956 a build-up at Wicken Fen resulted in a maximum of 50 birds by the 20th. In the period 1957-61 there were more records per annum than at any other time either before or since although nearly all of these involved single birds. Between 1964 and 1969 records became irregular and scarce. In the whole of this period there were about 50 records.

Figure 55. Snow Bunting – monthly distribution of all records.

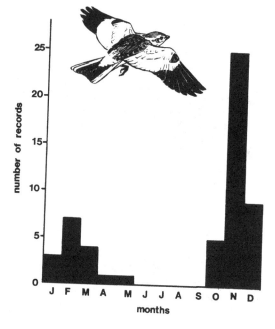

Present Status

An unusual and irregular winter visitor.

Apart from five records in 1970 there have been only single records in five other years in this period. An examination of the monthly distribution (Fig.55) shows a considerable bias towards November with most other months of the winter attracting equal numbers of records.

Earliest date: 13 October (1962 Heydon)

Latest date: 13 May (1977 Mepal)

Yellowhammer *Emberiza citrinella*

Pre 1934

Evans considered this species to be 'fairly common' and Lack stated that it was a common resident throughout the county.

1934-Present Status

A widespread and common resident.

Found throughout the county, thriving particularly in the area of the chalk uplands. This species is much more conspicuous in summer on arable farmland and along the hedgerows. A comparative study (Evans 1974) suggested that changes in agricultural practice in the Ely area had reduced the population of Yellowhammer and that this was partly the reason for the expansion into this sort of habitat by the Reed Bunting. After the initial insurgence of this latter species the situation seems to have stabilised and there is no sign of the Yellowhammer declining further. The figures from the Coton census area, however, seem to show the Reed Bunting increasing against a fairly stable population of Yellowhammer (Bucknell 1977). Morgan and O'Connor (1980) showed a direct relation between the density of this species and hedgerow length. There has been no infiltration of suburbia or urban districts and thus this species remains thoroughly a bird of open farmland. At many suitable (scrubland) sites small (20-50) roosts form and where food is concentrated flocks of 50 or more gather. An exceptionally large flock of 460 was noted at Brinkley in 1984.

Breeding Status

Generally widespread and fairly common but rather scattered in fenland. A count of 30 pairs along the Roman Road near the Gog Magog Hills was made in 1984. The *Breeding Atlas* (Sharrock) shows most squares occupied, but it would be surprising if the loss of hedgerow has not adversely-affected this species. On the Coton census site between 16 and 21 pairs bred (Bucknell 1977).

Ringing Results

The mainly sedentary nature of this species is indicated by the fact that only two ringing recoveries have been reported. The first was ringed at Dernford Fen, Sawston, in October 1977 and found at Knebworth (Herts), the following July. The second, showing movement in the opposite direction, was a bird ringed at Romford (Essex), in January 1960, which was found at Kirtling in August 1982.

EVANS P.J. A comparative study of the Yellowhammer and Reed Bunting. *Camb. Bird Club Report* 48. 1974.
MORGAN R.A. and O'CONNOR R.J. Farmland habitat and Yellowhammer distribution in Britain. *Bird Study* 27. 1980.

Cirl Bunting   *Emberiza cirlus*

Pre 1934

Evans thought it 'very rare' and Lack quoted two records: one in the summer of 1913 and another, nesting, in Cambridge in 1922. There was a further record of summering at Bury Fen on the Huntingdonshire border in 1927.

1934-1969

These records are listed by year in order to give the clearest picture of their interesting and unusual nature.

1943 One in the Grantchester/Cambridge area in late autumn.
1944 The species bred at Cambridge and at Orwell with possible breeding taking place at Lord's Bridge, Barton, Cambridge sewage farm and Westwick. Wintering records were received from Cambridge sewage Farm and Ely beet factory.

(This was part of a nationwide phenomenon suggesting an expansion of range or at least an invasion.)

1945 No breeding. Wintering was recorded as follows: a female at Cambridge sewage farm in January, a male and two females at Cottenham in February, a pair at Earith in February, and one at Stretham in March.
1948 One at Bottisham Park, 9 May.

1951 A male in the Botanic Garden, Cambridge, on 4 July; a male at Cambridge sewage farm, 14 August; and one at Cambridge sewage farm on 8 October.

1952 A male singing on the Gog Magog Hills, 27 April, and a male singing at a west Cambridgeshire site, 19 April.

1953 A pair probably bred on the Gog Magog Hills.

1954 A male at Coploe Hill, Ickleton, 7 August.

1955 Two at Hauxton, 13 February, and four on the Gog Magog Hills, 25 October.

1956 One at Over, 7 February.

1957 One at Milton, 7 November.

1959 A female at Balsham, 4 June, and a female at Fulbourn Fen, 1 November.

1961 An immaculate male between Orwell and Barrington, 20 February.

Present Status

A rare visitor.
A single record of a male at Dernford Fen, Sawston, on 20 April 1971. Following the extraordinary records of 1944 an attempt to establish a breeding population seems to have failed rapidly leaving only annual visitors, presumably from other English colonies. In recent years the range of this species has contracted alarmingly (Sitters 1985) and this might explain the paucity of records since 1970.

SITTERS H.P. Cirl Buntings in Britain. *Bird Study* 32. 1985.

Corn Buntings were the telegraph wire birds of South Cambridgeshire in the 1960s.

Reed Bunting   *Emberiza schoeniclus*

Pre 1934

Evans described this species as 'local'. Lack considered it fairly common although more numerous in summer than winter and noted interestingly that in winter it was often seen on agricultural land.

1934-Present Status

A common and widespread resident.

Found throughout the county but, as might be expected, its stronghold is in the fenland region and it is more scarce in the more built-up southern area. In the study conducted by Evans (1974) it emerged that the population had moved into arable regions largely as a result of the decline of the Yellowhammer which was under intense pressure from hedgerow removal. Evans postulated that there was not interspecific competition but showed a correlation between the population of both buntings with hedgerow length; the less hedgerow the more Reed Buntings and vice versa. This finding was confirmed on the Coton plot where Reed Bunting increased after hedgerow removal (see Yellowhammer). In some districts, in some conditions (usually harsh weather) these birds have entered gardens but in general this has not taken place on any regular basis. Roosting in winter is common although the size of each gathering varies considerably according to the time of the winter and the shelter opportunities; at

Figure 56. Reed Bunting – British recoveries of birds ringed in Cambridgeshire.

places such as Fulbourn and Wicken Fens and at Fowlmere from 20 to 250 can be found in due season.

Breeding Status
The *Breeding Atlas* (Sharrock) shows this species to be breeding in each 10 km square with probably a greater density in the northern, wetter districts. On the Ouse Washes counts revealed 200-500 pairs in 1972, 850 pairs in 1975, 320 pairs in 1982 and 346 in 1984. On the Coton census site numbers varied from 2 to 12 (Bucknell 1977).

Ringing Results
There have been no foreign recoveries, all movement being within England and mostly south for the winter. The recoveries are shown in Figure 56.

EVANS P.J. A comparative study of the Yellowhammer and Reed Bunting. *Camb. Bird Club Report* 48. 1974.

Corn Bunting *Miliaria calandra*

Pre 1934
Evans considered it to be 'abundant'. Lack, however, thought it was of a rather scattered distribution although breeding throughout the county.

1934-Present Status
A moderately common resident.
Found particularly in the arable areas where any hedgerow persists, and tolerably common along the wide uncultivated droves in parts of open fenland. Like the other buntings, but more so, this species is no lover of suburbia and is therefore under increasing pressure from both the agricultural changes and lack of alternative habitat. At a number of sites, particularly those surrounded by farmland (Fowlmere watercress beds, Fulbourn and Wicken Fens), roosts form with quite large numbers at times (50-500) and these birds commonly arrive at roost sites an hour or so before sunset and can be seen fliting from bush to bush.

Breeding Status
A common breeding species in the southern agricultural areas and quite common in areas of fenland. The *Breeding Atlas* (Sharrock) shows a very scattered and patchy distribution in Cambridgeshire, but this is true also of most other counties. On the Coton census site numbers were fairly stable throughout the ten years, varying between 2 and 11 pairs.

APPENDIX – LOCAL ORGANISATIONS

THE CAMBRIDGE BIRD CLUB

The club was formed in 1925 by W.H. Thorpe, Bernard Tucker, L.J. Turtle, the Marquis Hachisuka and J.D. Clarke as a frustrated response to the blocking of a move to form different sections within the Cambridge Natural History Society. Originally the membership was restricted to 20 but within five years this was relaxed. The first report was published in 1927. It was originally called the Cambridge Ornithological Club but was changed to the Cambridge Bird Club in 1930. The club has been responsible for keeping records of the county avifauna ever since and has grown considerably to around 350 to 400 members. The Executive Council is responsible for its day to day running and consists of a Chairman, Treasurer, Town Secretary and Undergraduate Secretary, and six other members. Among its officers are: the County Recorder, the Report Editor, the Research Officer, a Field Meetings Secretary and a Ringing Secretary. Meetings are held fortnightly in the lecture rooms of the Departments of Applied Biology and Zoology throughout the September to May period. Field trips by coach and car are arranged within East Anglia (and sometimes further) to sites of ornithological interest. An emphasis on scientific recording and research is matched by the organised work parties to local sites, particularly Wicken Fen, Minsmere and Adam's Road Bird Sanctuary. The annual subscription at the time of writing is £5 and a list of the officers of the club can be found in the Cambridge City Library in the Lion Yard.

CAMBRIDGESHIRE WILDLIFE TRUST

CWT was founded (as the Cambridgeshire and Isle of Ely Naturalists' Trust) in 1956 in line with the other county naturalists' trusts and its aim was stated as being;

To record and study the chief places of natural history interest in Cambridgeshire and the Isle of Ely.

To protect these places if they are threatened.

To acquire and administer any such place as a Local Nature Reserve if this action is the most appropriate method for conservation.

To co-operate with other local and national bodies with interests in natural history and nature conservation.

To encourage interest and understanding for an intelligent policy of nature conservation, which should not run counter to the best interests of agriculture, forestry, sport, and other rural industries and occupations.

The running of the trust is similar in concept to that of the Bird Club with the exception that the trust has a full-time professional staff to execute the decisions of its executive council. The trust is also part of the Royal Society for Nature Conservation which serves as an umbrella organisation for all the county trusts. CWT has a membership of around 2700 at the time of writing (1987) and a subscription of £9. Its number of reserves is increasing and stands at 27 at present. In recent years the trust has become thoroughly involved with the protection of many threatened sites in the county and is in great need of voluntary help of all kinds. It also operates a shop within the office selling many items of interest. Contact can be made via the telephone on Cambridge 358144 and the Trust Office is at 5 Fulbourn Manor, Fulbourn, Cambs.

ROYAL SOCIETY FOR THE PROTECTION OF BIRDS LOCAL MEMBERS GROUP

Formed in 1977, the group followed the emerging pattern of the RSPB to provide more for the practical members of the society. Meetings, which are widely advertised, are held in winter on Tuesdays in the University Chemical Laboratory, Lensfield Road, Cambridge and are mostly based on travels of birdwatchers or experiences of RSPB staff and wardens. Some organised visits and work parties are arranged. There are many areas of overlap with the Cambridge Bird Club providing double the opportunities that most communities have.

At the time of writing the group leader is: John Owen who can be contacted on Cambridge 60648.

INDEX

of English names of birds in the text